I0484559

UNIVERSITY OF OKLAHOMA

GRADUATE COLLEGE

UNIQUE *ANSA*-BIS(INDENYL)METALLOCENE DICHLORIDES:

THE QUEST FOR CATALYSTS

A Dissertation

SUBMITTED TO THE GRADUATE FACULTY

in partial fulfillment of the requirements for the

degree of

DOCTOR OF PHILOSOPHY

By

RAYMOND LAVERN FRAZIER II

Norman, Oklahoma

2003

UMI Number: 3109072

INFORMATION TO USERS

The quality of this reproduction is dependent upon the quality of the copy submitted. Broken or indistinct print, colored or poor quality illustrations and photographs, print bleed-through, substandard margins, and improper alignment can adversely affect reproduction.

In the unlikely event that the author did not send a complete manuscript and there are missing pages, these will be noted. Also, if unauthorized copyright material had to be removed, a note will indicate the deletion.

UNIQUE *(ANSA)*-BIS(INDENYL) METALLOCENE DICHLORIDES:
THE QUEST FOR CATALYSTS

A Dissertation APPROVED FOR THE
UNIVERSITY OF OKLAHOMA
DEPARTMENT OF CHEMISTRY AND BIOCHEMISTRY

BY

R. L. Halterman, Ph.D.

C. Leroy Blank, Ph.D.

Daniel T. Glatzhofer, Ph.D.

Lance L. Lobban, Ph.D.

Kenneth M. Nicholas, Ph.D.

ACKNOWLEDGEMENTS

Obviously a dissertation thesis is not created in a vacuum. My family, and especially my children, Ray and Victoria, have suffered through the long process of research with its long laboratory hours and the resultant absence of my presence at home, as well as the deteriorated temper that such stress engenders. Thank you for putting up with a "grumpy" Daddy for so long.

Dr. Ronald Halterman is a research advisor that can only be described as stupendous. His great breadth of knowledge in the field of Organic and Organometallic chemistry is second only to the depth of his compassion and endurance in putting up with an older, less traditional student with familial commitments that often distracted from the work in the lab and certainly slowed the process. I can only hope that someday I am as able to lend a compassionate ear to my students as he is to his, for I know my understanding of Organic Chemistry will never equal his. My thanks for setting my role model in teaching demeanor as well as for broadening my own understanding of the field.

To all the past and present members of Group Halterman for paving the way and helping me keep my sanity, my thanks, especially to Jason Shipman, Alexander Tretyakov, and David Combs whose theses contributed much of the background for my own. Special thanks to Eric Bailly, my friend and lab mate, for teaching me the "way of the lab" and to Lisa Crow, my friend and extra ear, for letting me bounce ideas off her and listening to me ramble for the last five and a half years.

I also wish to thank the office staff, especially Arlene and Theresa, without whose aid graduate school would have been much more difficult, if not impossible, and the other

iv

organic faculty; especially Dr. Nelson, without whose cumes and verbal support therein I probably would not have survived to the third year; Dr. Schmitz, now retired, whose presence and never failing kind word in the hall always makes one feel welcome—a sure sign of true class; Dr. Glatzhofer, whose willingness to listen and advise during the absences of Dr. Halterman made those times more bearable.

Finally, I would like to thank my committee members, Drs. Blank, Glatzhofer, Halterman, Lobban and Nicholas, for the willingness to serve on my committee; the time they put into my various meetings; the endurance with which they put up with my questions over the years; and the effort which they put into correcting mistakes that would otherwise have found their way into this thesis. To each of you, my sincere thanks and appreciation.

TABLE OF CONTENTS

LIST OF FIGURES

ix

x

xi

xiii

xiv

xv

LIST OF TABLES

UNIQUE (*ANSA*)-BIS(INDENYL) METALLOCENE DICHLORIDES:

THE QUEST FOR CATALYSTS

ABSTRACT

Ethylene-bridged-1,2-bis(2-menthylinden-1-yl)zirconium and -titanium dichlorides have been synthesized from *ansa*-1,2-bis(2-menthylinden-1-yl)ethane **79**. The known *ansa*-1,2-bis(2-bromoinden-1-yl)ethane reacted with the organozinc reagent formed from menthylmagnesium chloride and zinc chloride in the presence of palladium in order to provide ligand **79**. The dilithium salt of this ligand was formed and reacted with zirconium(IV)chloride and titanium(III)chloride to give the respective metallocenes.

Uniquely tethered metallocene propylene-bridged-1-(4-methylinden-7-yl)-3-(inden-1-yl)zirconium dichloride and ethylene-bridged-1-(4-methylinden-7-yl)-2-(inden-2-yl)zirconium dichloride were also synthesized. The propylene-bridged zirconocene has significant mobility in its bridging unit which allows it to interconvert between conformers rapidly. The ethylene-bridged zirconocene forms enantiomers. The formation of these two metallocenes is due in large part to the methodology of combining a Stetter Reaction and a Erker Cyclization to generate a 3-(4-methylinden-7-yl)propanoic acid from a starting diacid, ketoglutaric acid.

The background, methodology, design, and synthesis of these four metallocenes will be discussed. In addition, other less successful attempts will be discussed as regards limits to this new methodology.

xviii

Chapter 1

Modern Uses of Zirconocenes

1.1   Introduction

While Group 4 metallocene complexes of bis(cyclopentadienyl)-,bis(indenyl)-, bis(fluroenyl)- and combinations thereof, were developed primarily as achiral or racemic polymerization catalysts, their synthetic use in modern times is steadily increasing as they find application in many other reactions [1]. The vast array of reactions to which metallocenes, and especially zirconocenes, are being applied includes Diels-Alder reactions [2], alkene hydrogenations [3], ketone and imine reductions [4], C-F bond activations [5], carbomagnesations and carboaluminations [6], and cyclizations of compounds containing two or more pi systems [7]. The majority of these reactions utilize a single enantiomer of a zirconocene to induce asymmetry in the products, providing regio-, stereo-, enantio-and/or diastereoselective transformations. Among the great multitude of these reactions, those which exhibit enantioselectivity are, perhaps, the most useful.

While many different metallocenes can be used for some of these reactions, the focus here will be primarily upon zirconocenes. In fact, the emphasis will be even more specific than just zirconocenes for the emphasis will be upon chiral *ansa*-zirconocenes.

1

Ansa, Greek for handle, implies that the metal center of the zirconocene is tethered with a bidentate ligand, the handle usually being a short carbon or heteroatomic chain connecting the two binding sites of the bidentate ligand. The chirality of the metallocene may be due to chiral substituents on the ligand or to chirality derived from the metal's binding to specific "faces" of a planar ligand such as a cyclopentadienyl moiety, termed planar chirality. This planar chirality will be discussed in greater detail later, though it follows essentially the same rules as "normal" chirality (see section 2.4) by assuming the metal equally bound to each carbon to generate p-*R* and p-*S* configurations. Metal-centered chirality is also possible but less useful in asymmetric transformations, as the general mechanism for these reactions involves ligand dissociation and could thus involve racemization if an achiral intermediate formed.

In the ligand-derived chiral metallocenes with cyclopentadienyl moieties, the cyclopentadienyl moieties bind strongly to the metal; other ligands, usually halide or methyl, dissociate and the metallocene's chirality is unaffected during reaction and thus imparts chirality to products, often in an enantioselective (one enantiomer preferred over the other) fashion. The actual degree of selectivity is measured by the enantiomeric excess (% ee) which is the difference between the enantiomers divided by the sum of the enantiomers expressed as a per cent.

2

1.2   Modern Enantioselective Uses of Chiral Zirconocenes

The Diels-Alder reaction is one of the reactions in which zirconocenes are finding

use as catalysts for asymmetric transformations. For example, the Diels-Alder reaction

between oxazolidinone-based dienophiles, such as the oxazolidinone **1,** and

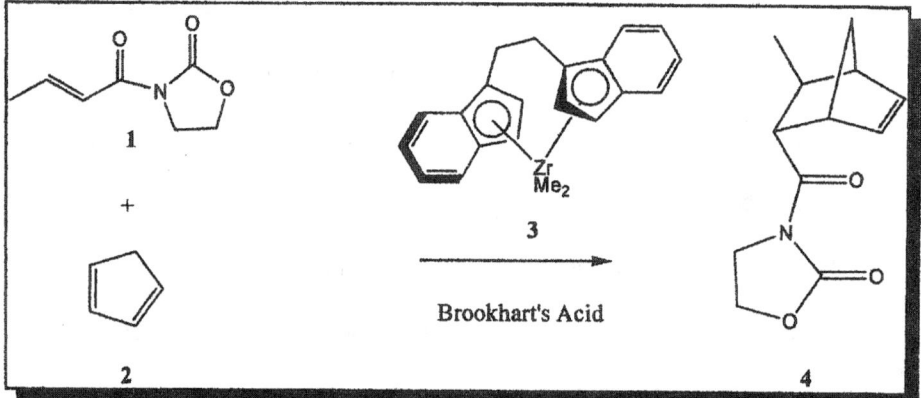

Figure 1 Asymmetric Diels-Alder Reaction

cyclopentadiene (**2**) can be catalyzed with the single enantiomer of Brintzinger's *ansa*-

ethylene-bridged bis-indenyl complex (**3**) [(S,S)-1,2-bis(inden-1-yl) ethane] in the

presence of Brookhart's Acid {[(Et$_2$O)$_2$H][B(C$_6$H$_3$CF$_3$CF$_3$)$_4$]} to yield adduct **4** in up to

95% enantiomeric excess, depending on solvent choice and catalyst loading [2a].

The mechanism of this reaction involves two molecules of **1** replacing the methyl

groups in the oxidized catalyst [2a]. Once the two molecules are bound to the zirconium

center, one of the conjugated dienophiles is activated to cyclopentadiene's attack.

However, the zirconocene does not just activate the dienophile but also directs the

3

incoming cyclopentadiene to a specific "face" of the diene by occluding the other (as seen

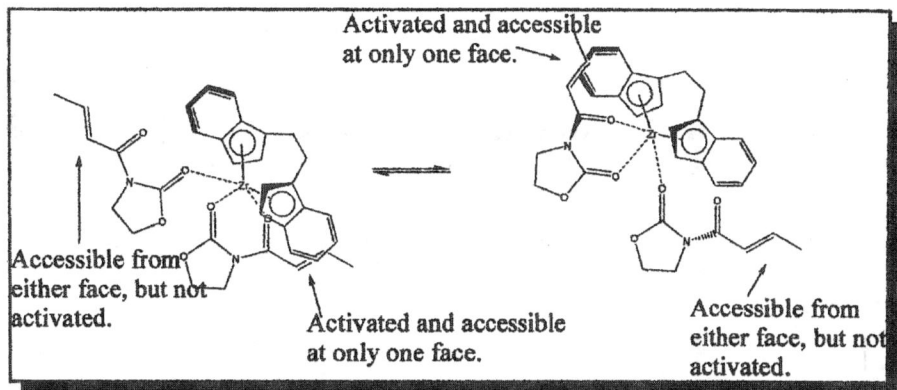

Figure 2 Stereoorientation of Enantioselective Diels-Alder Reaction

in **Figure 2**). This stereodirecting effect, along with the typical Diels-Alder π-overlap,

accounts for the high degree of enantioselectivity.

This use of a zirconocene to catalyze a reaction to an enantioselective or

enantiospecific product is not, of course, unique to Diels-Alder reactions. Among alkene

hydrogenations, chiral zirconocenes can similarly catalyze the reduction of the alkene to a

specific enantiomer. Troutman, Apella, and Buchwald have used (*R,R*)- or (*S,S*)-

(EBTHI)ZrMe$_2$ in conjunction with dimethylphenylammonium

tetra(pentafluorophenyl)borate salts, to generate catalysts that can reduce even tetra-

substituted olefins with a high degree of enantioselectivity [3b]. One of the many such

reactions studied by these authors is the reduction of 2,3- dimethylindene (**5**) to the

corresponding (S,S)-1,2-dimethylindane (**6**). This hydrogenation reaction, shown in

Figure 3, is catalyzed by the (R,R)-metallocene in 87% yield with a 93% enantiomeric

excess[3b]. An investigation of this mechanism [3b] revealed that the great degree of

4

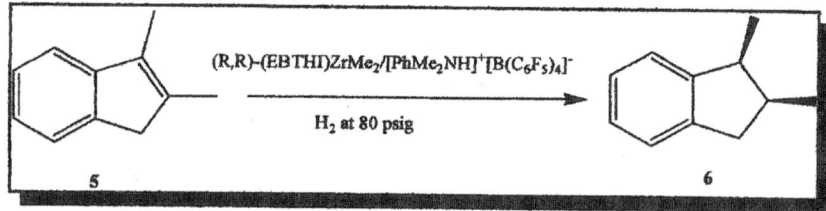

Figure 3 Asymmetric Hydrogenation

enantioselectivity is due to the fact that only one face of the indenyl species may approach

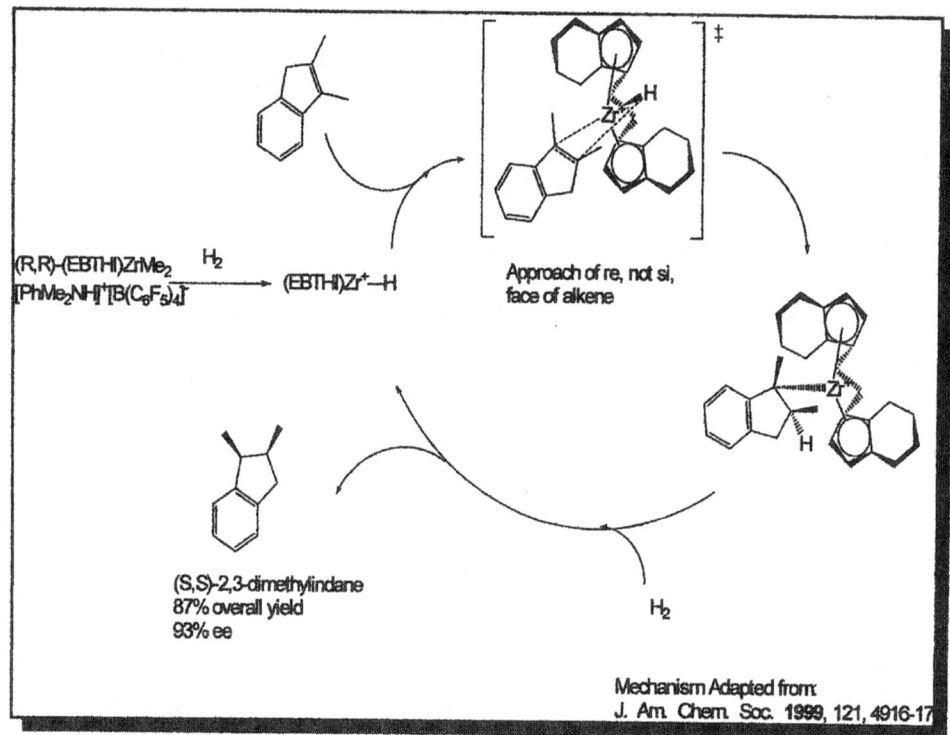

Figure 4 Mechanism of Asymmetric Hydrogenation

the metal center. As may be seen in **Figure 4**, the face that becomes the a (*S,S*)-indane

(here, the re face) can approach the zirconium center while the face that would have

resulted in the (*R,R*)-indane is prevented due to steric interactions. This interaction with

5

the catalytic center results in a highly enantioselective hydrogenation.

1.3   A Modern Regioselective Use of Zirconocenes

Besides enantioselective hydrogenation, another useful type of reduction

catalyzed by zirconocene compounds is the regioselective reduction of the double bond

between carbon and a heteroatom such as oxygen (ketones) or nitrogen (imines). An

example of this type of reduction can be found in the work of Ishii et al, wherein ß-

hydroxy ketones, such as **7**, reacted with aldehydes in the presence of bis-(cyclo-

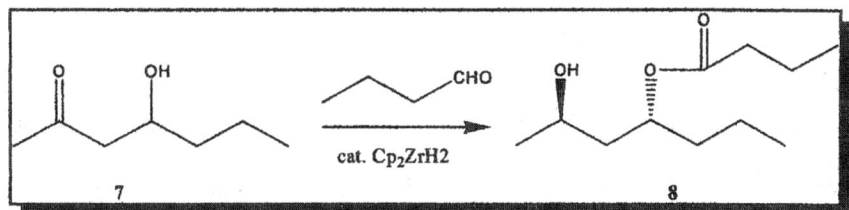

Figure 5 Regiospecific Reduction of β-Hydroxy ketones

pentadienyl)zirconium(IV)dihydride, (as well as the corresponding hydrochloride and

dichloride zirconocenes) to stereoselectively produce monoesters of the corresponding

anti-1,3-diols, such as **8** [4a]. The immediate importance of this reaction may be lost if

one is not reminded that this product is only a trans-esterification away from the anti-

diols that are "difficult to prepare by conventional methods" [4a].

Indeed, conventional reduction methods, such as sodium borohydride or lithium

6

aluminum hydride in the presence of a chelating metal, leads to a high preference for the

syn diols. An example of this typical preference is shown in Bonini's work [157] as

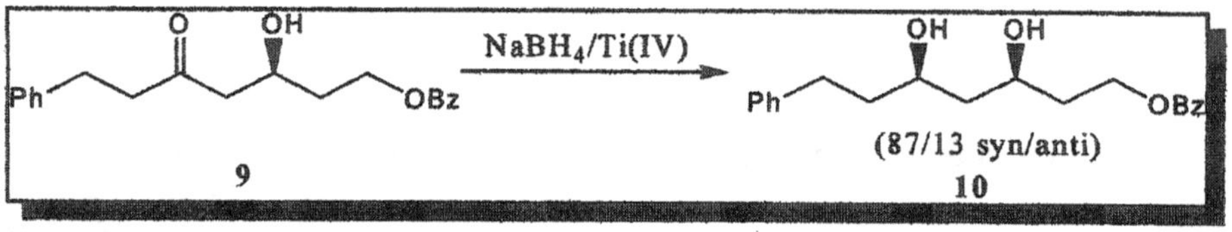

Figure 6 Typical Syn Diol Formation

shown in **Figure 6** wherein alpha-hydroxy ketone **9** is converted to diol **10** in an 87/13

syn- to anti- ratio.

1.4   A Novel Use of Zirconocenes

Besides the regio- and enantioselective reactions that metallocenes catalyze, these

reagents are also useful for performing reactions that are difficult to achieve any other

way. One novel example of this type of reaction is the activation of carbon-fluorine

bonds. The carbon-fluorine bond is exceedingly strong and therefore difficult to activate.

In fact, it is even more difficult to activate than the carbon-hydrogen bond of alkanes

[5b].

However, recent work by Edelbach et al. has shown that $[Cp_2ZrH_2]_2$ will activate

the C-F bond of hexafluorobenzene to form $Cp_2Zr(C_6F_5)F$ (**Compound 11**) [5a]. This

fluoride may then be transferred to another species with the lost fluoride being replaced

7

with a hydride, as in **Compound 12**. If the process could be made catalytic and the cycle

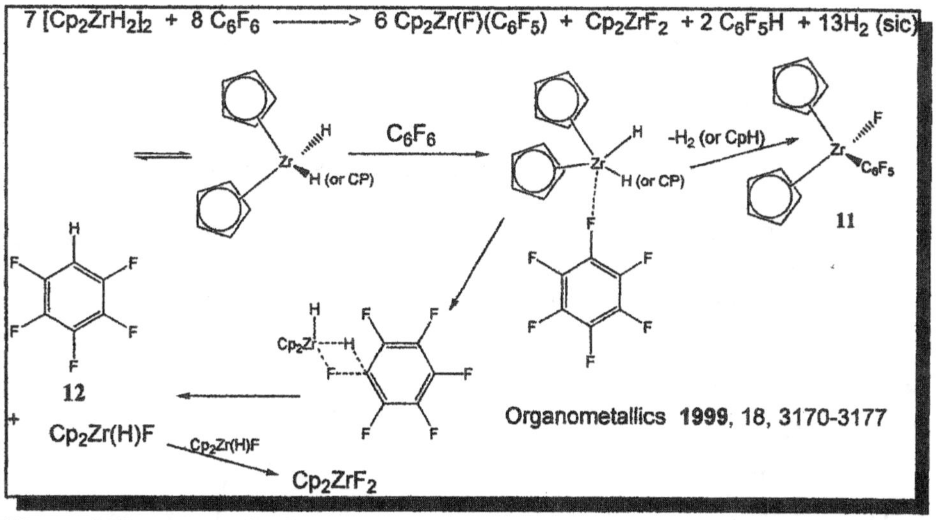

$$7 \, [Cp_2ZrH_2]_2 \; + \; 8 \, C_6F_6 \longrightarrow 6 \, Cp_2Zr(F)(C_6F_5) \; + \; Cp_2ZrF_2 \; + \; 2 \, C_6F_5H \; + \; 13H_2 \; (\text{sic})$$

Organometallics **1999**, 18, 3170-3177

Figure 7 Fluorine Activation

repeated successively on the same molecule, the possibility exists to eventually de-

fluorinate fluorinated compounds. The potential for practical applications in being able

to transfer a fluoride ion in the manner of a hydride from the $Cp_2Zr(C_6F_5)F$, or possible

defluorination of fluorinated hydrocarbons with further developments, speaks for itself.

1.5 Modern Uses of Zirconocenes to Effect Mild Transformations

As already seen, metallocenes allow stereochemical control of products in typical

reactions, rendering them enantioselective and diastereoselective, and simplification of

extremely difficult reactions, as in defluorination. Metallocenes are also being used for

milder, more selective introduction of functional groups, especially to pi systems, than is

8

traditionally observed.

An example of this use for metallocenes to effect mild, selective introduction of functional groups is found in conjunction with carbometalations, such as carbomagnesations (also called carbomagnesiations) and carboaluminations. These carbometalations were first reported by Dzhemilev [8]. In these reactions, a Grignard-like species is added across a double-bond [14]. An example of carbomagnesiation is shown in **Figure 8**, wherein ethyl magnesium chloride in added across the double bond in

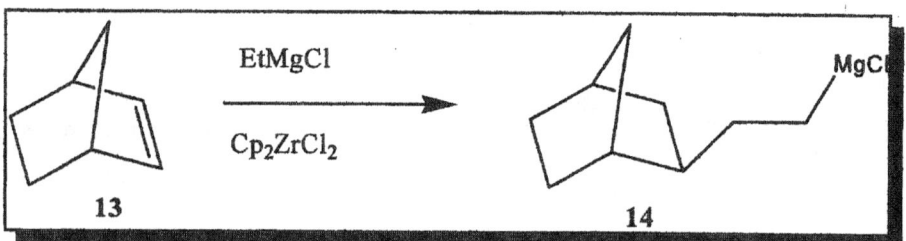

Figure 8 Typical Carbomagnesiation

compound **13** to give compound **14**.

By the early 1990's Hoveyda and others had fine-tuned the carbomagnesation reactions to be highly regio- and stereoselective [9-14]. By 1995, Hoveyda was able to add enantio- and diastereoselective to the list [10]. By the end of the 1990's, carboaluminations were also under study for enantioselectivity via the chiral metallocenes [11].

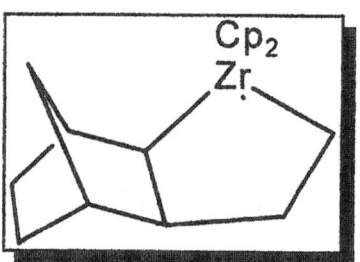

Figure 9 Zirconacyclopentane

9

Some of the early mechanistic studies performed by Negishi [12] and Waymouth [13] shed light on the zirconacyclopentane intermediate (**Figure 9**). However, it was Hoveyda that first proposed the complicated pathway involving bis(zirconocene) complex

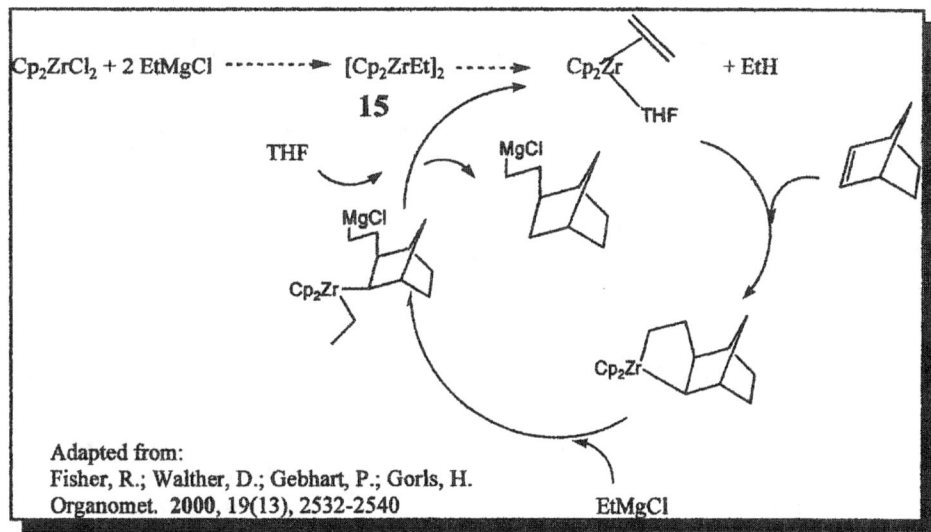

Figure 10 Catalytic Carbomagnesation Mechanism

15 for the carbomagnesation mechanism (**Figure 10**) which eventually prevailed [14]. Quite recently, Gebhardt and co-workers were able to isolate intermediates of this more complicated catalytic pathway, including the bis(zirconocene) complex 15 [15].

Though many more types of metallocene-assisted reactions already exist, and more are constantly being developed, the only other one to be mentioned here is the cyclization of diolefins. Especially in the field of natural product synthesis where the regio-, stereo-, and enantioselectivity of rings is of the utmost importance, and mild conditions a distinct boon, these cyclization reactions are finding great use. In point of fact, the cyclizations of the diolefins are often, though not always, accomplished via the

10

routes of carboaluminations (to the corresponding aluminacycle and conversion to the cyclic organic compound) [16] and carbomagnesations [17] (through the corresponding magnesacycle).

The generalized mechanism for these reactions depends rather on the type of π-system, as shown in the two general mechanisms proposed by Trost [17] for terminal alkenes and alkynes (**Figure 11**). However, the similarity to the carbomagnesation mechanism becomes obvious when

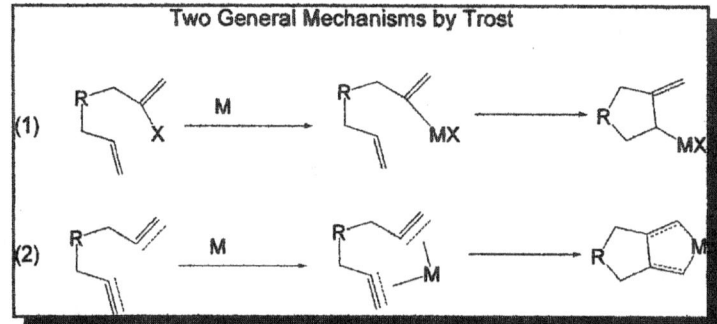

Figure 11 General Mechanisms by Trost [17]

Waymouth's proposed mechanism (**Figure 12**) [16] for the cyclization is examined.

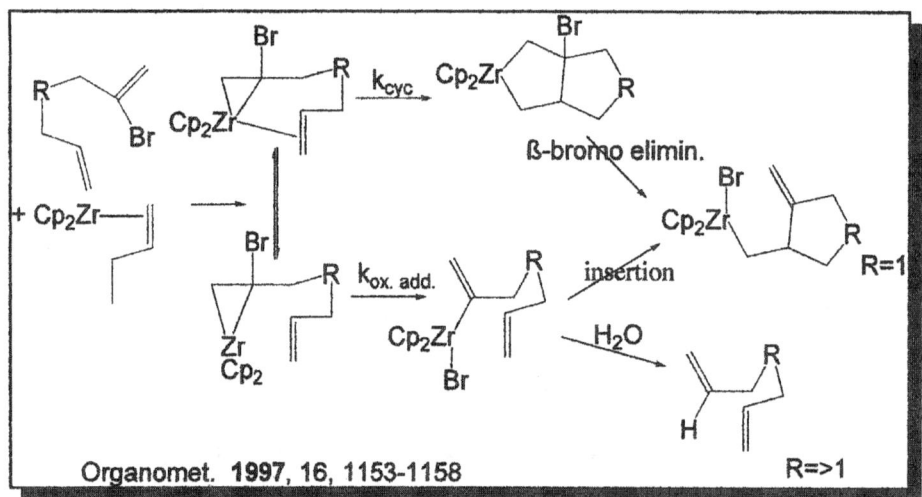

Figure 12 Carbomagnesation Mechanism by Waymouth [16]

11

However, not all zirconocene-assisted cyclizations are done via carbometalation (magnesation or alumination) [18]. For example, Wender, Rice and Schnute were able to take advantage of a zirconocene-mediated enyne cyclization to formally synthesize enantiomerically pure phorbol (**Figure 13**),

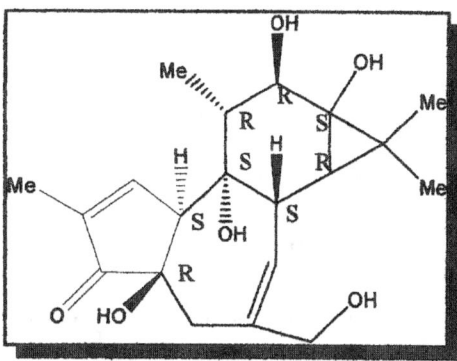

Figure 13 Natural Phorbol [(+)-Phorbol]

a goal that had eluded researchers since its racemic synthesis in 1989 [19].

In their synthesis, no aluminum or magnesium is present, so the cyclization obviously does not go through an aluminacycle or magnesacycle. It is likely that the mechanism involves an oxidative addition of both pi systems to the zirconocene followed

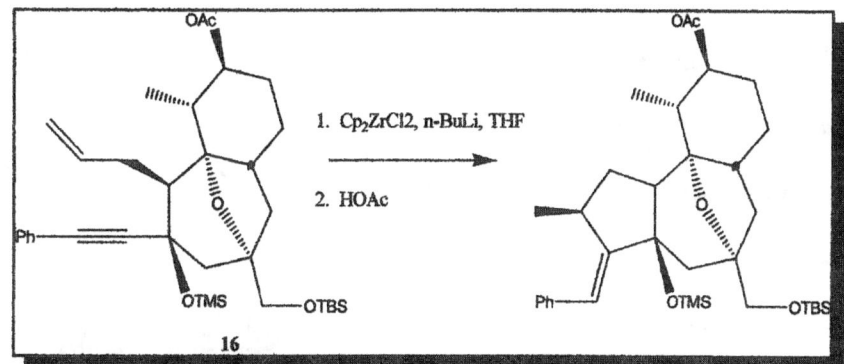

Figure 14 Enyne Cyclization in Route to (+)-Phorbol

by reductive elimination to give the desired, optically active product (+)-phorbol from enyne **16** (**Figure 14**).

12

1.6 The Need for Asymmetric Synthesis in Modern Pharmaceuticals

This synthesis of the single enantiomer of natural (+)-phorbol (**Figure 15**) illustrates both the power and need for asymmetric reactions. Natural products and biochemically active species are generally found as single enantiomers rather than as racemates. In order to mimic nature in an effort to better understand her, single enantiomers of natural and biochemical products must be accessible. Asymmetric

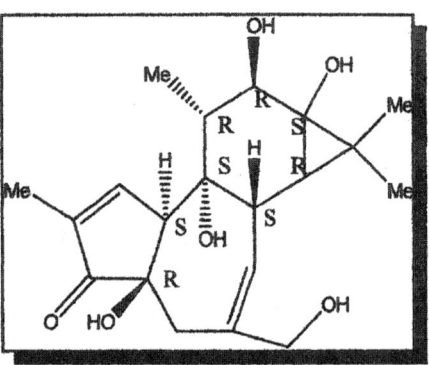

Figure 15 (+)-Phorbol

synthesis, such as that of (+)-phorbol, demonstrates a powerful method for such access.

In addition to making single enantiomers in order to better understand nature, it is also necessary to be able to make single enantiomers to better interact with her, as nature often recognizes only one of a pair of enantiomers. For example, the amino acid alanine is always found in nature as the (S)-alanine. The (R)-alanine (unnatural alanine) is unrecognized by the body's enzymes and therefore may not be metabolized [159].

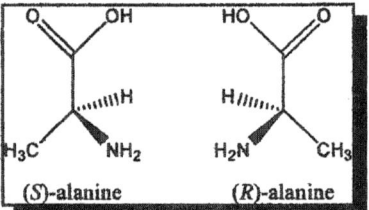

Figure 16 Alanines

Similarly, in drug design, one enantiomer of a drug usually interacts with the receptor site or sites of an enzyme while the other enantiomer does not, rather like a right

13

hand fitting a right glove while the left hand does not. While there are cases in which both enantiomers of a drug can fit into a receptor site, it is usually a looser fit and a less active drug, rather like the loose fit of reversible gloves that fit either hand, but slip off too easily to be used in detailed work. In other words, the best drugs interact enantiospecifically with the receptor [163].

An example of enantiospecific drug interaction is found in the common analgesic Ibuprofen (**Figure 17**). Only the (S)-Ibuprofen is active while the (R)-Ibuprofen is inactive [160]. Therefore, only half as large a dose of the single enantiomer drug is needed as compared to the racemic form. This increased value, of activity to dose, also allows pharmaceutical companies to patent a single enantiomer drug when only the racemic form has previously been known, translating into new patents and big money for the company able to develop (and market) the pure active enantiomer.

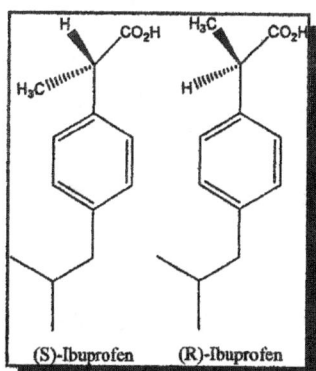

Figure 17 (R)- and (S)-Ibuprofen

To further complicate issues, sometimes each enantiomer of a drug interacts with different enzymes in different ways. For example, (S)-(E)-6-(bromoethylene)-3-(1-naphthalenyl)-2H-tetrahydropyran-2-one [(S)-BEL] is ten times more selective for the iPLA(2)beta enzyme than the

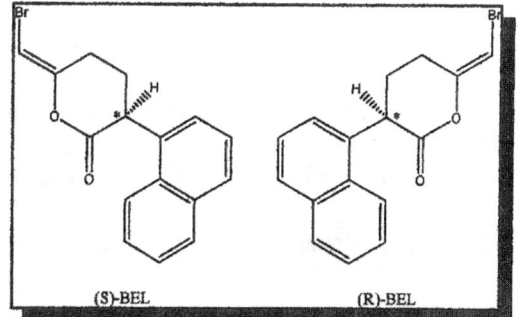

Figure 18 (S)- and (R)- BEL

14

iPLA(2)gamma enzyme while the reverse is true for (*R*)-BEL [161]. This difference of interaction may be beneficial as a medical probe, allowing selective inhibition of different enzymes as in the above case, but it can also be detrimental as in the classic case of thalidomide whose *R*-enantiomer acts as a mild sedative and whose S-enantiomer acts as a powerful teratogen [162].

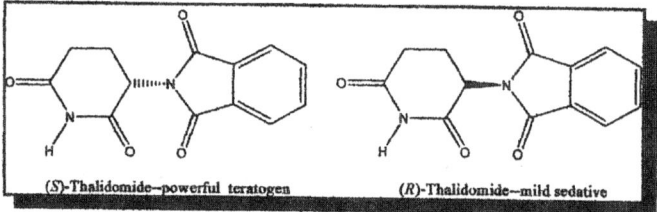

Figure 19 Enantiomers with very different activities

Based on the detrimental effects of the medically-inactive enantiomer in some drugs (most notably: thalidomide, penicillamine, ethambutol [163]) originally sold as racemates (equal quantities of both enantiomers) the Federal Drug Administration moved in January of 2003 to switch "to the pure enantiomer for older drugs and will only approve single isomers of new chiral drugs" [163].

While currently single enantiomer drugs make up only 38% of the global pharmaceutical market, the percentage has risen steadily since 1997 [164] and will obviously climb even higher with the FDA's current movement. It is therefore extremely important in medical research and pharmaceutical research that single enantiomers of chiral compounds be accessible. Since chiral metallocenes allow the incorporation of chirality to previously racemic or pro-chiral compounds in a catalytic manner with a high degree of enantioselectivity, they are a good choice for the chirality-inducing step in multiple step synthesis towards single enantiomer compounds.

15

1.7   Conclusion

In conclusion of this overview of the state of modern zirconocenes, the great utility of these catalysts, especially in regards to selectivity issues, has, hopefully, been demonstrated. Reactions not normally regio-, stereo-, enantio-, or diastereoselective can at times be made so by the addition of a catalytic amount of the appropriate zirconocene. As previously mentioned, for the modern synthetic chemist, enantioselectivity is probably the most important of these, especially in medical and pharmaceutical research. Through chiral zirconocenes such as the Brintzinger catalyst, Diels-Alder reactions [2c], carbomagnesations [10], carboaluminations [11], and enyne [19] and nitrile [7a] cyclizations have been rendered enantioselective. Other zirconocene catalytic systems have also been used to render alkene hydrogenations [3c], ketone [4a] and imine reductions [4c], and many other such reactions, enantioselective (predominantly one enantiomer over the other) or even enantiospecific (exclusively one enantiomer over the other). Other reactions, such as C-F bond activation, that are unavailable by more traditional means, are available via zirconocene activation. Finally, the use of chiral zirconocenes is opening mild reaction pathways, even cyclization routes, for the asymmetric synthesis of previously elusive chiral compounds, as in the case of (+)-phorbol [19]. Applications for metallocenes are already multitudinous and are continuing to grow on a daily basis, especially in medical and pharmaceutical research. With the FDA's current pressure to have all new chiral drugs formed as single enantiomers, the use of chiral zirconocenes is sure to see an even greater surge over the next few years.

16

Chapter 2

Development of Zirconocene Catalysts for Polymerization

2.1 Introduction

While much of the current research in the field of metallocenes involves using single enantiomers of chiral metallocenes to asymmetrically catalyze transformations to yield selective, asymmetric products, as seen in the first chapter, the impressive development of zirconocenes as catalysts received its primary impetus, as previously mentioned, with the polymerization of olefins. In point of fact, polymerization still dominates the research ends of a large number of those investigating zirconocenes, as it did in the early years of zirconocene history. A complete history of zirconium itself takes one back to its discovery in 1789 by Klaproth [20a]. Klaproth discovered the metal shortly before the 1791 discovery of titanium by William Gregor [20b] and 35 years before J. J. Berzelius isolated the metal in pure enough form for its entry into industrial use [20a] for what would, nowadays, be called: material science applications. The history of zirconocenes as catalysts, however, begins with the history of polymerizations, for which they were eventually developed.

17

2.2   Early Polymerizations

In the pioneering days of polymerization, 1900-1953, it was impossible to "polymerize olefins to very high molecular weight chains, let alone stereoregular polymers" [21]. Indeed, it was this very inability to synthesize polymers that led Wolfgang Ostwald to coin the phrase "land of neglected dimensions" in 1917 [22]. Until that time, there were only a few known natural polymers, such as rubber (known forever and vulcanized by Goodyear in 1839 [21]) and nitrocellulose (discovered in 1838 [21]), and only one known synthetic plastic, a phenol-formaldehyde resin synthesized by Baekland in 1907 [23] and subsequently used as a varnish. Not until the 1930's did a synthetic rubber, polystyrene-butadiene, reach production (1930 in Germany, 1933 in the US) [23].

In the in-between years of the 1920's, little progress was made in the actual polymerizations of compounds, but much was done in the methodology for the study of such compounds [21]. In 1930, a major development for polymerization itself occurred, though it would not be capitalized for many years, Friedrich and Marvel discovered the first polymerization of ethylene in the presence of lithium alkyls [24]. Later, a low density polyethylene (LDPE) was discovered when ethylene was subjected to high temperature and pressure in the presence of benzaldehyde [25]. For most of the next two decades, to the start of the 1950's, various groups continued attempts at polymerization of olefins, vinyl ethers, vinyl halides and other such potential monomeric species via cationic, anionic and free radical processes [21]. The most successful polymerization

18

attempts at this time, those obtaining polymers of significant molecular weight, used Friedel-Crafts-type catalysts [26-29,31].

The highest reported molecular weight of this time period was from a group at the Socony-Vacuum Labs with high molecular weights of polypropylenes that reached 880,000 units [27]. One of the pioneers of the field who was able to consistently obtain high molecular weight polymers was Schildknecht. In 1947, he had polymerized vinyl ethers to a molecular weight of 810,000 units with alkylated aluminum-promoted and metal cation-catalyzed conditions [28]. A year later he reported polymerizing vinyl isobutyl ether with BF_3 to get tacky, but high molecular weight polymers [29]. He was able to get a non-tacky, crystalline version by running the same reaction with 1% BF3-ether at -80 to -60 °C. The difference between the two he attributed to differences in structures between polymer chains, one he called an alternating *d,l* and the other repeating *d*, or *l* [29]. Natta suggested applying a new terminology for these types of polymers, isotactic for those having the same orientation and syndiotactic for those having alternating orientations [30]. Schildknecht also screened various transition metals as polymerization catalysts, with alkyl aluminums, the monomers they catalyzed, and a range of temperatures for polymerization [31].

While Schildknecht's work was undoubtably important and ground-breaking for the type of polymerizations that Ziegler would later make famous [32], his work was with vinyl ethers, not the olefins that were the "Holy Grail" of polymerization attempts at the time. It was Karl Ziegler, in the aftermath of World War II, who continued the work broken off by Friederich and Marvel [24] and developed the modern field of polymers.

19

2.3   Ziegler-Natta Catalysts for Polymerization

By the early 1950's, Karl Ziegler had been publishing in the field of olefin polymerization for more than twenty years [33, 34]. Until 1950, his work had primarily centered on various radical-initiated polymerizations [34-40 and references therein]. Thereafter, he returned to the anionic routes of Friederich and Marvel. He had briefly investigated these lithium alkyl-promoted polymerizations [41-45] in the mid 1930's before dismissing them as forming polymers of too low a molecular weight to be useful [24]. In attempting to "tweak" the anionic polymerizations to higher molecular weights, Ziegler and Gellert realized that the primary problem with the low weights was the polymer eliminating and precipitating lithium hydride out of solution [24].

While changing the solvent from hydrocarbon to ether did help somewhat, the real advance came when they tried a relatively new, ether-soluble compound, lithium aluminum hydride [24]. The trialkyl aluminum formed when the lithium aluminum hydride reacted with the olefin (especially ethylene) was found to polymerize the olefin even better than did the ethyl lithium that they were trying to prepare [24, 46, 47].

At this point, Ziegler obviously knew he was on to something, in the next few years he patented over a hundred and fifty variations on this reaction with lithium aluminum hydrides, trialkylaluminums, triarylaluminums, as well as Ga, In, Be, and Mg versions of the same [48]. At this point, Ziegler began mixing combinations of the metals, such as magnesium or zinc with aluminum halides or alkylaluminums and trying different states of ethylene, such as hot, gaseous instead of cold under pressure[48f] in a manner

20

very reminiscent of Schildknect.

Among the frenzied activity of various projects in Ziegler's lab, were several that would prove to be of immense importance. The first of these occurred in 1953 when Ziegler and Holzkamp were running a polymerization reaction that produced only 1-butene due to some nickel contaminants left in the reactor [49]. This fortunate finding showed that the termination step of the polymerization, the reason for only lower molecular weight polyethylenes (usually less than 30, 000), could be controlled [49]. With Holzkamp continuing to work on the nickel, Ziegler put Breil onto screening a series of transition metals [49] with results very similar to Schildknect's screening of vinyl ethers.

While cobalt and platinum compounds worked like nickel, when ethylene reacted with trialkyl aluminum in the presence of zirconium acetylacetonate, high molecular weight linear polyethylenes were produced even at low pressures [49]. With this result, so amazing that it was ignored the first time crystalline (rather than amorphous) polymers were found in the reactor [49], their search scanned through much of the periodic table, their best results always being with salts of Groups IV-VI, and especially titanium(IV) chloride and aluminum(III) chloride which Ziegler designated the "Mülheim Atmospheric Polyethylene Process"[48h, 49]. With this process, linear, crystalline polymers with molecular weights up to three million were now readily produced [50].

The second of the developments in Ziegler's lab that would eventually have far reaching effects was a new methodology to prepare sodium and lithium cyclopentadienyls [51]. However, while this new methodology for the cyclopentadienyl anion was

21

developed in Ziegler's lab, very little development of the methodology occurred there. Indeed, even his polymerization did not see full fruition in his lab, as his lab was concentrated primarily on ethylene polymerizations [49]. However, while the full value of Ziegler's catalysts was discovered elsewhere, his ambition and enthusiasm truly launched the field of olefin polymerization. By 1956 when Karl Ziegler wrote the review of the new organometallic field, he was able to quote forty-three of his own papers (out of the sixty references given) [52].

The full potential of Ziegler's catalyst was discovered and developed in detail by Giulio Natta. Ziegler had solved the problem of reaching linear, high molecular weight polymers. However, the industrially sponsored institute for which he worked, the Max Planck Institute for Coal Research, was primarily interested in polyethylenes and copolymers of ethylene, so Ziegler "shared" his results with the Montecatini Company and the Goodrich-Gulf Chemical Company [49]. Giulio Natta was a consultant with the Montecatini Company and had been involved previously in kinetic studies of aluminum alkyls with olefins [49]. When the company reached a deal with Ziegler, he quickly went to work with the Mülheim catalyst[49].

Applying some of his earlier results, Natta reduced the titanium(IV) tetrachloride to titanium (III) trichloride and was able to quickly obtain some of the most crystalline polymers ever produced, as well as the very first crystalline polypropylene, poly-α-butylene and polystyrene [49, 53]. Besides modifying the Mülheim catalyst to improve crystalinity and applying it to compounds that had yet to be polymerized (other than in co-polymerizations with ethylene), Natta also reflected back to work of Schildknecht and

22

his amorphous polymers and crystalline polymers.

As previously mentioned, Schildknecht's theory was that the difference between types of polymers was due to orientations of what is now called the pendant group. Natta, using x-ray techniques, was able to determine that the principal chain of his crystalline polymers contained stretches of three units, not two as would have been the case if the configuration were alternating *d* and *l* as Schildknecht had suggested [53]. Since there is an odd number of repeating units, to get the regularity of the crystal, Natta realized they must be oriented in the same direction, and, therefore, Schildknecht's polymers were reversed. He suggested terming this type of polymer with all orientations in the same direction, as was previously mentioned, isotactic [53]. Later that same year, 1955, Natta did discover crystalline polymers with their carbons "asymmetrically arranged in regular sequence," which he defined as syndiotactic [54]. This discovery was found in a homogeneous catalytic system such as Schildknect used, so the dismissal of Schildknect's assignment of the orientations in the polyvinyl ethers may have been premature. However, be that as it may, over the next several years, the blossoming field of polymer chemistry came into full bloom as the attempted optimization of the Mülheim-type catalytic system yielded more and more information as to how it works and how to produce certain types of polymers.

For their work in discovering and developing the Mülheim-type catalysts that have come to be known as Ziegler-Natta catalysts, Ziegler and Natta were awarded the 1963 Nobel Prize in Chemistry. In his Nobel Prize presentation speech, Professor A. Fredga defines the Ziegler Catalyst as a "combination of aluminum compounds with other

23

metallic compounds . . . which can be used to control polymerizations and to obtain

molecular chains of the required length [for synthetic utility] [55]." He goes on to applaud

Natta for finding:

> "stereoregular macromolecules, i.e. macromolecules with a
>
> spatially uniform structure . . . [in which] all the side groups point
>
> to the right or to the left, these chains being called isotactic. [. . .]
>
> The secret [of which] is that the molecular environment of the
>
> metal atom, at which new units are stuck onto the chain . . . , is so
>
> shaped that it permits only a definite orientation of the side groups
>
> [55]."

The "molecular environment" Fredga mentions changes with different Ziegler-

Natta catalysts. This may result in different stereospecific polymer structures being

formed, dependent on the choice of the catalyst and starting olefin [56]. In modern times,

it is fairly well established that, at least for propylene, "the energy difference between

stereomeric forms of the metallocene catalyst determines the elastomeric property of the

final polypropylene" [57]. It is also a good general rule that symmetry correlates to

tacticity, C_2-symmetrical zirconocenes forming isotactic polypropylene, for example [87].

24

2.4 Early Development of Ziegler-Natta Catalytic Systems

In the earlier days of Ziegler-Natta catalysts, trial and error, not energy calculations or symmetry-design, led to the discovery of new metallocenes and the type of polymer produced. Over the first few decades after the development of Ziegler-Natta catalysts, much research was devoted to the improvement of the system by modifying the "environment of the metal atom" of which Fredga spoke [55]. Modifications were quickly discovered that allowed the formation of linear isotactic, syndiotactic, and atactic polymers, which was the primary goal of the trial and error research of those next few decades.

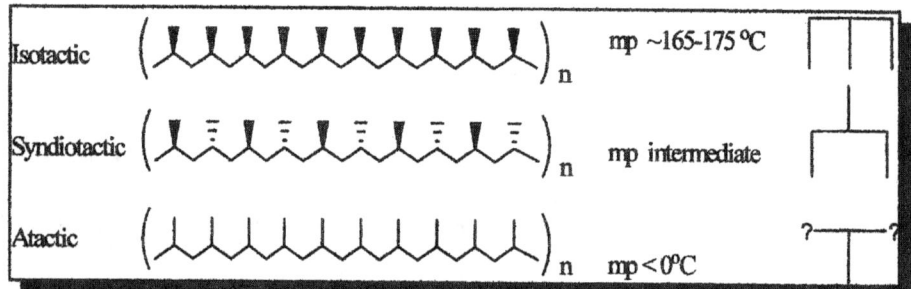

Figure 1 Isotactic, Syndiotactic and Atactic Polymers

One of the modifications to the system that was to have lasting significance was the introduction of alkyl groups, especially cyclopentadienyl, to the transition metal catalyst to aid in its solubility in preparation of homogeneous catalysts. In the early days of the field of Ziegler-Natta polymerizations (1950's and 60's), the cyclopentadienes added were simple, seldom substituted, and, if substituted, only mono-substituted [58].

While some earlier work with cyclopentadienyl ligands was done on ferrocene and

25

manganese complexes [58], it is worth noting some of the "firsts" of cyclopentadienyls and related compounds in regards to the Ziegler-Natta type catalysts: The first application of cyclopentadiene to a Group IV metal is found in a 1953 patent application for a process that produces "dibutoxycylopentadienyltitanium chloride" and "dicyclopentadienyltitanium chloride" [60]. (As a point of note, the "bis" nomenclature so familiar to today's chemist's seems to have first been applied in 1958, but was not used with frequency until much later [64].) The first utilization of fluorene and indene as a ligand to metals of Group IV, V, VI is Archibald P. Stuart's use in 1954 [61]. The first journal article published citing cyclopentadienyl as a ligand to a Group IV metal is Birmingham, Seyferth, and Wilkinson's 1954 article, wherein they form dicyclopentadienyl-titanium dichloride [62]. The first use of cyclopentadiene to form an actual zirconocene complex was accomplished by David S. Breslow in 1954 to form a "di(cyclopentadienyl)zirconium salt" [63].

The number of the above "firsts" demonstrates the immediate fervor of activity generated by Ziegler-Natta catalysts with cyclopentadienyl moiety ligands. It might be noted that these cyclopentadienyl moieties, whether cyclopentadienes, indenes, fluorenes, or substituted versions thereof, were generally produced with the same cyclopentadienyl moiety as both the first and second ligand on the metal center, which is to say, achirally. It was not until the 1970's that the issue of chirality would be added to these systems [58].

26

2.5   Basic Issues of Chirality

The formal study of chirality itself dates back to 1848 and Louis Pasteur's separation of (+)-tartaric acid from (-)-tartaric acid and the observation that each pure form rotated the plane of polarized light an equal amount in opposite directions [65]. Pasteur proved that optical rotation was a molecular property and theorized that molecular dissymmetry was the cause [65].   More than a quarter of a century later the cause of that dissymmetry, spatial arrangement around the carbon atom of four different groups in three dimensional space, was proposed by Van't Hoff and J. A. Le Bel [66]. The term chiral, Greek for handed, was applied when it was noted that the two possible forms are non-superimposable mirror images, as are human hands. While the angle between carbons is 109.5° not the 90° that simply twisting a square would give, the basic

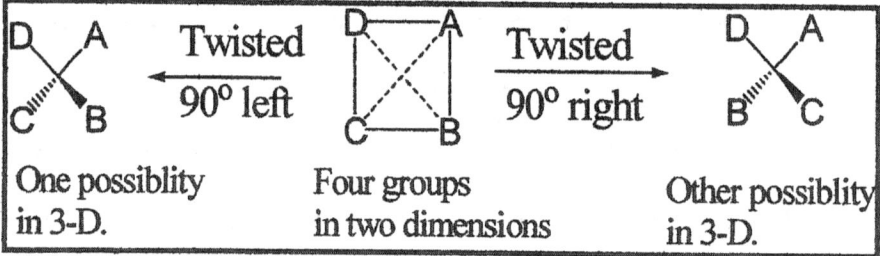

Figure 2 Chirality in a Nutshell

concept in Figure 2 is correct. For his efforts in proving his theory, against significant opposition by the established scientific community, Van't Hoff was named the first recipient of the Nobel Prize in 1901 [66].

The two possibilities for the arrangement of four different groups are termed

27

enantiomers. One of a pair of enantiomers turns plane polarized light clockwise and is designated (+) or (d), while the other turns it counterclockwise and is designated (-) or (l). An equal amount of each enantiomer will cancel each other's optical rotation and is termed a racemic mix or racemates. In order to distinguish between enantiomers in literature without having to do optical rotation measurements, a nomenclature system (called the Cahn-Ingold-Prelog convention) was developed in which the four groups are prioritized by atomic mass of the element to which the central group is directly bound (ties move to the next element, π-bonds are treated as if each bond was to a separate atom) and the lowest priority group is placed farthest away from the viewer. Then, the increasing priorities of the remaining groups are examined. If the priorities increase clockwise, they are assigned an (R) [rectus], if counterclockwise an (S) [sinister].

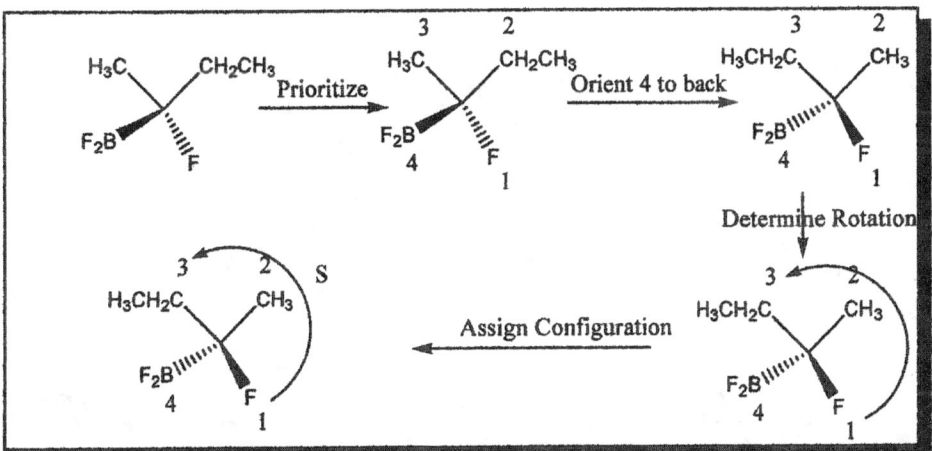

Figure 3 *R/S* Assignment

It is worth noting that optical rotation [(+) and (-)] is experimentally determined

28

while R and S configurations are arbitrarily assigned. Therefore, there is no direct relation between the two. Indirectly, however, if (+) is known to be (R) for a particular pair of enantiomers, then (-) must be (S), but only for that particular enantiomeric pair.

Especially in organic chemistry, compounds may be quite complex. So far, only molecules with a single chiral carbon have been discussed. However, every carbon that has four different groups attached is a chiral or stereogenic carbon. If a molecule has only one such stereogenic carbon, the molecule is chiral and two enantiomers exist.

However, if a molecule has two or more stereogenic carbons, the molecule may or may not be chiral. If the molecule is non-superimposable on its mirror image, it is chiral. If each and every stereogenic center of a chiral molecule is inverted, R changed to S and vice versa, then the enantiomer is formed. Enantiomers have the same physical

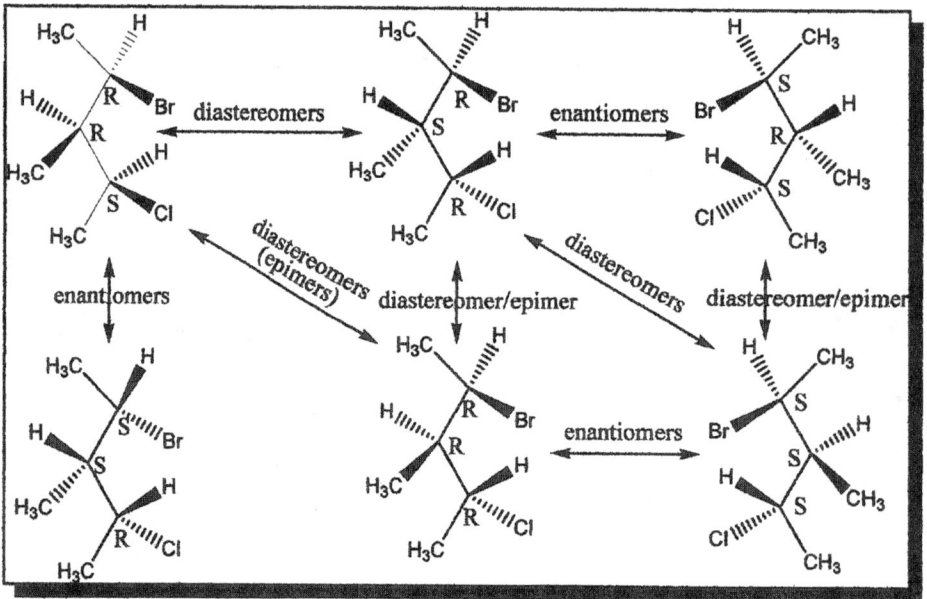

Figure 4 Stereomeric Relationships with Multiple Stereogenic Centers

29

properties. If not every center is inverted, then diastereomers are formed. If only one stereogenic center is inverted, then a type of diastereomer termed an epimer is formed. Unlike enantiomers, diastereomers and epimers have different physical properties.

On the other hand, if a molecule with two or more stereogenic carbons is superimposable on its mirror image, then it is achiral. The most common case of superimposable mirror images is due to a mirror plane of symmetry in the molecule. This mirror image is in fact the same compound. This type of molecule is called meso.

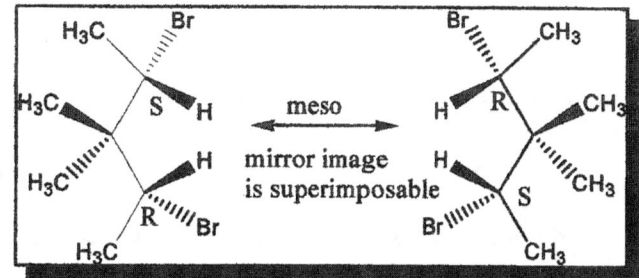

Figure 5 Superimposable Mirror Image is Meso

2.6 Chiral Metals and Metallocenes

By 1900, Pope and Peachey had extended the concept of chirality to metals with their announcement of an optically

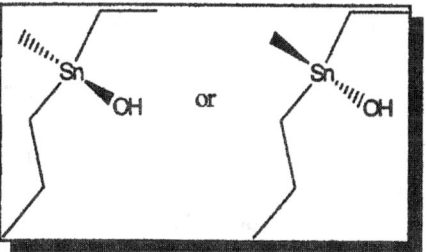

Figure 6 First Reported Chiral Metal

active methylethylpropyltin hydroxide (**Figure 6**) [67]. This metal-centered chirality [68] was introduced to Ziegler-Natta catalysts with cyclopentadienyl ligands in 1974 as a reference for "studies on dynamic stereochemistry around the Ti atom" [58, 69]. The first chiral zirconocenes with a cyclopentadienyl moiety were described in 1975 with the

30

production of $(\eta^5\text{-}C_5H_5)(\eta^5\text{-}C_5H_4CHMe_2)ZrClR$ where R = cyclohexyl, $PhCH_2O$, or $PhCH_2$ [70]. These first chiral versions were chiral due to four different substituents about the metal center and were also formed racemically. However, the following year Leblanc and Moise reported synthesizing "enantiomerically enriched monosubstituted cyclopentadiene"[58] which when metalated, allowed the isolation of optically active titanocenes[59, 71]. The degree of preference for one enantiomer over the other is generally measured in enantiomeric excess (e.e.), defined as the difference between the enantiomers divided by the total of both enantiomers, multiplied by 100%. If the optical rotation of one of the pure enantiomers is known, the other must be the same value in the opposite direction, so the measure of the optical purity gives the enantiomeric excess.

When optically active metallocenes were applied to propylene polymerizations, it was found that the "chirality of the catalytic system and the conformation of the monomeric unit of the growing isotactic chain bound to the catalyst controlled enantioface discrimination of the monomer and thus the stereospecificity" [158]. In other words, in isotactic propylene polymerization, if one enantiomer of the metallocene gave the pendant methyls on one side of the growing chain, the other enantiomer would give them on the opposite side. This exciting result spurred the search for other chiral catalytic systems.

In catalytic systems, metal-centered chirality, wherein four separate groups are tetrahedrally arranged around a metal center, is, as previously mentioned, generally less useful than ligand-derived chirality. The most obvious method for generating a chiral ligand is to attach a chiral substituent to a good ligand, such as a cyclopentadienyl. The

31

cheapest and easiest chiral substituents are derived from nature.

One of the first compounds generated in this burgeoning field was also the first annulated chiral cyclopentadiene (**Figure 7**). This

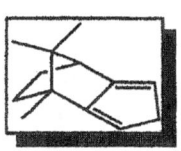

compound was synthesized by Burgstahler in 1976 [59] from camphor, but with yields too low to be useful. It was not until nearly a decade later that two useful synthetic routes to Burgstahler's annulated cyclopentadiene were published, one by Halterman and Vollhardt and one by Paquette, McKinney, and McLaughlin [59]. Such ligands that can form diastereomeric metallocenes based on the face of the ligand that binds to the metal are called diastereotopic ligands. This particular ligand, due to the methyl protruding over one face, is selectively metallated to a single diastereomer. Thus, by starting with the naturally pure, optically active (+)-camphor and generating a

preferred diastereomer of the metallocene, a single enantiomer of the metallocene can be preferentially generated in a fairly facile manner as shown in **Figure 8**.

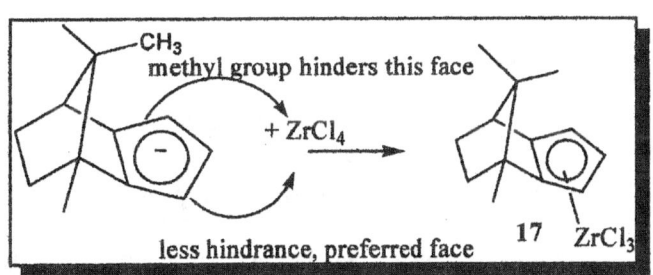

Figure 8 Unhindered Face Attacks to Preferred Diastereomer

With methodology to form chiral cyclopentadienyl compounds (and the chiral metallocenes therefrom) now readily available, knowledge of chiral metallocenes expanded quickly. One of those expanding this burgeoning field was Kagan and his collaborators. In 1978, Kagan and his team substituted cyclopentadiene via naturally-

32

occurring (-)-menthol derivatives to form (-)-menthylcyclopentadiene and (+)-neomenthylcyclopentadiene [59], shown in **Figure 9.** The single sigma (σ) bond between the cyclopentadiene and the chiral substituent in these ligands makes the faces of the cyclopentadienyl anion equivalent in the metallation. Such ligands that can result in only one metallocene regardless of the face metalated are called homotopic ligands.

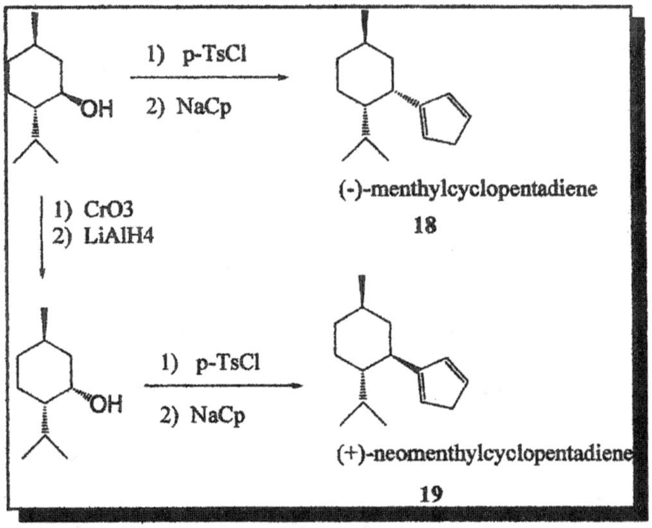

Figure 9 Kagan's Natural Chiral Substituents

Besides naturally chiral substituents, Kagan also explored the qualities necessary to make good chiral ligands. These qualities, as outlined by Kagan [95], include staying coordinated to the metal during the reaction, having catalytic activity comparable to the achiral version, being easily modifiable to allow variations, simple to prepare, and preferably able to synthesize either enantiomer at will. Since cyclopentadiene and its moieties (indenyl, fluorenyl, and substituted versions thereof) generally have these qualities, it is no wonder that their use became standard in the development of chiral metallocenes for polymerizations (and other asymmetric reactions).

Over the next few decades, the search for more such catalytic systems containing

33

chiral cyclopentadienyl-type metallocenes flourished. For example, a more substituted version of Kagan's (+)-neomenthylcyclopentadiene compound was prepared a decade later by Halterman and Vollhardt from menthone (**Compound 20**) by Michael addition of the phenyl Grignard followed by a stereoselective reduction of the resulting ketone[59]. The alcohol formed by this reduction (**Compound 21**) was mesylated then displaced with sodium cyclopentadiene to yield the phenyl-substituted version, **Compound 22**.

Halterman and Vollhardt also published the first C2-symmetrical annulated

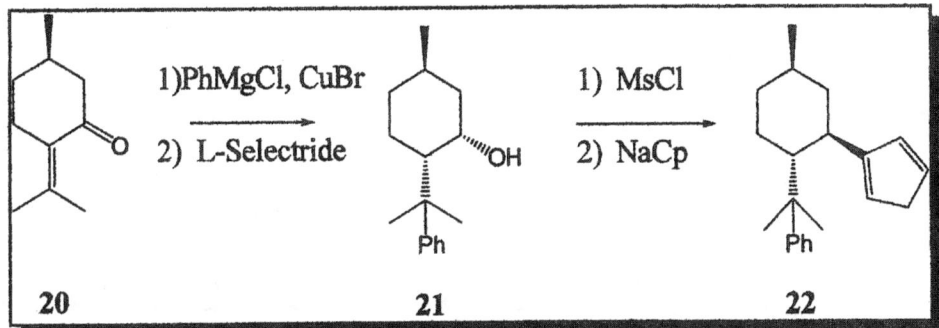

Figure 10 Halterman and Vollhardt Substituted Version

cyclopentadiene compound (**Figure 11**) in 1986 [59]. This C2-symmetry allows for a homotopic ligand in the more rigid annulated system, rather than the usual diastereotopic ligand. While only homotopic and diastereotopic ligands have been discussed thus far, enantiotopic ligands also exist. An excellent review of such chiral cyclopentadienyl compounds and their differences is found in Halterman's 1992 *Chemical Reviews* article [59].

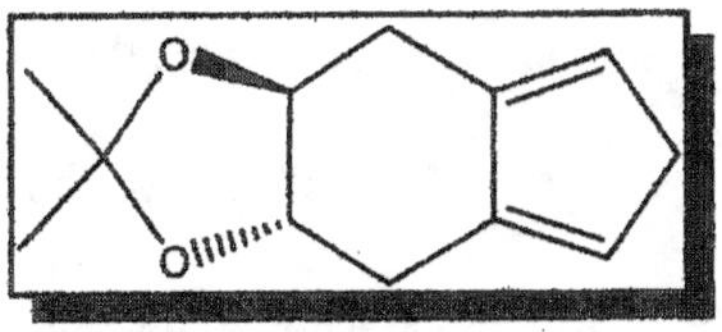

Figure 11 C$_2$ Symmetry

34

The earliest examples of enantiotopic cyclopentadienyl ligands are disubstituted cyclopentadienes [59]. While they are not themselves chiral, they form chiral metallocenes on metalation, and are, therefore, referred to as prochiral. For example, 1-ethyl-2-methylcyclopentadiene [59] could be deprotonated and metalated to give the chiral racemic mixture of metallocenes **23a** and **23b** shown in Figure **12**. It might be noted here that by convention, the planar cyclopentadienyl moiety is assumed to be in the plane of the board while the metal, which is bound approximately equally to each of the five carbons in the ring, is shown to be coming straight to the viewer by a line eclipsing the lines of the ring or straight away from the viewer by a line eclipsed by the lines of the ring.

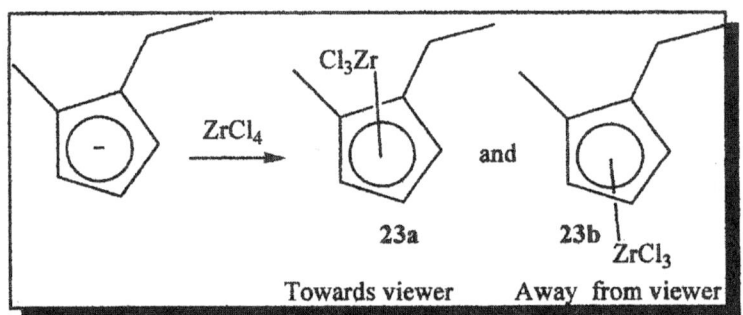

Figure 12 Enantiomers of Metallocene **23**

This type of chiral metallocene is a little unusual in terms of its chirality. Usually, chiral molecules have a tetrahedral carbon with four different substituents. These carbons, however, are still planar. The metal binds firmly to the π electrons, though, and so once bound to a face, the metal remains on the face to which it is bound. This results in a pair of metallocenes who are non-superimposable mirror images of each other, i.e. enantiomers. Despite the planar nature of the carbons, the metallocene is chiral. This type of chirality is termed planar chirality and is fairly common among metallocenes

35

made from cyclopentadienyl moieties.

Up to this point, for demonstration purposes, a single displacement of halide from a metal tetrachloride by cyclopentadienyl moieties has been shown. However, in reality, usually two halides are actually displaced in formation of the metallocene (titanocene or zirconocene) to form bis(cyclopentadienyl)-type metal dichlorides. Taking advantage of this fact, in 1978 Brintzinger bridged two cyclopentadienyl compounds to form a bidentate type of ligand. This ligand was metallated to form *ansa*-trimethylenebis(η5-1-(3-tert-butylcyclopentadienyl) titanium (IV) dichloride [59], the first ansa-(Latin for bent handle)-metallocene (shown in **Figure 13**). The synthesis of this C2-symmetrical bridged complex initiated an intensive search for similar chiral *ansa*-metallocenes and new synthetic routes to them [59]. Many new compounds of this nature were

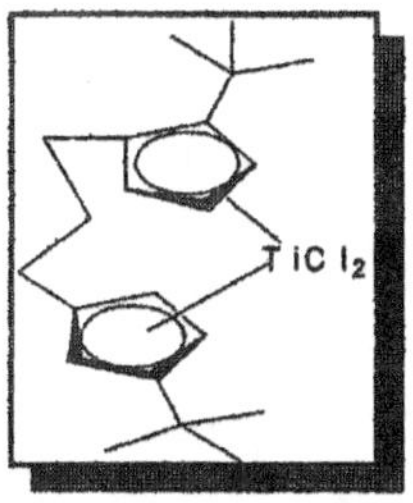

Figure 13 First
ansa-Metallocene

discovered by Brintzinger and other groups, as well as new synthetic strategies to the same [74].

One of Brintzinger's new compounds that was to have great ramifications was ethylene-1,2-bis(3-indene)zirconium(IV) dichloride [59] and its hydrogenated form, ethylene-bis(4,5,6,7-tetrahydro-1-indenyl)zirconium(IV) dichloride [59,75]. The novelty of this ligand, besides providing a homogeneous catalyst when metalated, lies in its

36

prochiral nature. Depending on which faces of the "didentate" ligand the metal binds, an RR (1R or pR), SS (1S or pS), or RS (meso) isomer is formed. The p's represent planar chirality. The nomenclature of the isomers formed is determined by the stereochemistry of the 1-position, generally considered the bridgehead in Brintzinger-type systems, and the assumption that the metal is equally bound to all the cyclopentadienyl carbons, exemplified here in **Figures 14 and 15** by the isomers of Brintzinger's ethylene-bridged bis(indenyl) zirconium dichloride.

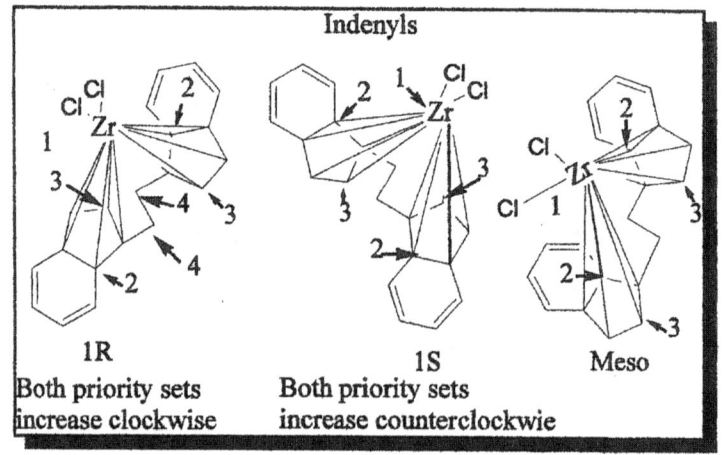

Figure 14 Brintzinger's Ethylene Bridged bis(Indenyl) Zirconocenes

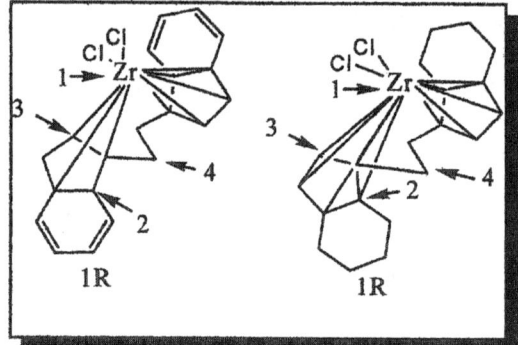

Figure 15 Priorities for Naming Brintzinger's EBI and EBTHI as 1R

37

2.7 Third Generation Catalysts

Chiral Group IV Metallocenes as a growing field received a boost in the late 1970's when it was discovered that these complexes could asymmetrically catalyze the hydrogenation of alkenes to form specific enantiomers [58]. For example, 2-phenyl-1-butene could be hydrogenated in the presence of homogeneous catalysts of (-)-menthylcyclopentadienyl-titanocenes to give (R)-(-)-2-phenylbutane or (+)-neomenthylcyclopentadienyltitanocenes to give (S)-(-)-2-phenylbutane[72]. This discovery helped initiate the field of catalytic asymmetric reactions which was briefly summarized in the first chapter, and provided an impetus for the discovery of new single enantiomer metallocenes.

However, the biggest boost to the field came in a combination of H. H. Brintzinger's work with bridged bis(cyclopentadienyls) and W. Kaminsky's fortuitously broken NMR tube (reminiscent of Ziegler's Ni-contaminated reactor vessel) which allowed methylaluminoxane (MAO) formation which caused a "lazy catalyst [bis(cyclopentadienyl)dimethyltitanium(IV)] dormant in the NMR tube [to] suddenly [become] sensationally active [73]." When Kaminsky's discovery of MAO's co-catalytic activity was applied to Brintzinger's complex, EBTHI Zr(IV) dichloride, new heights of polymerization activity were reached [76]. This combination also allowed the first production of chiral polymers [77]. With this first production of chiral oligiomers, the classical field of polymers came to an end and the modern era began. As Kaminsky and Sinn state in the Preface to <u>Transition Metals and Organometallics as Catalysts for Olefin</u>

38

Polymerization:

"first generation Ziegler-Natta catalysts . . . displayed only minor activities of about 30 kg polyethylene per g titanium . . . compared to . . . second generation catalysts, i.e. supported catalysts consisting of magnesiumdichloride or other magnesium compounds, titaniumtetrachloride, and aluminiumtrialkyls, [in which] the activities could be raised by a factor of 100. New impulses for olefin polymerizations have been brought about by the discovery of the very high co-catalytic activity of alumoxanes instead of aluminiumalkyls together with the soluble transition metal compounds of titanocene, zirconocene and hafnocene [78]."

These new impulses are still the driving force of metallocene research today in what has come to be called the "third generation of catalysts [79]." Homogeneous catalytic systems, generally chiral Group IV metallocenes containing various cyclopentadienyl moieties, allow for chemo-, regio-, and enantio-selectivity not found in hetereogeneous systems [80]. Homogeneous systems are also much easier to explain and understand at the molecular level, due to easier research conditions [81], especially as regards mechanistic studies [56]. Perhaps this explains why in 1988 Kaminsky, at the edge of the modern era of catalysis, states that "Although to a great extent there is a technical application for these catalysts, up to now the nature of the active centres and many reaction mechanisms are not completely known [78]."

39

This is not to say nothing was known of mechanisms at this time. Much research, by many groups, into mechanisms, had been done. One such research group was that led by Henrici-Olivée. It was Henrici-Olivée, in 1981, using a soluble bis-cyclopentadienyl titanium(IV) dichloride (**Figure 16**), who concretely determined the transition metal to be the active site [56] of polymerization [82].

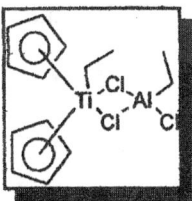

Figure 16 Active Species

In the two decades since Henrici-Olivée made his original discoveries, many other details of the mechanism of alkene polymerization have been

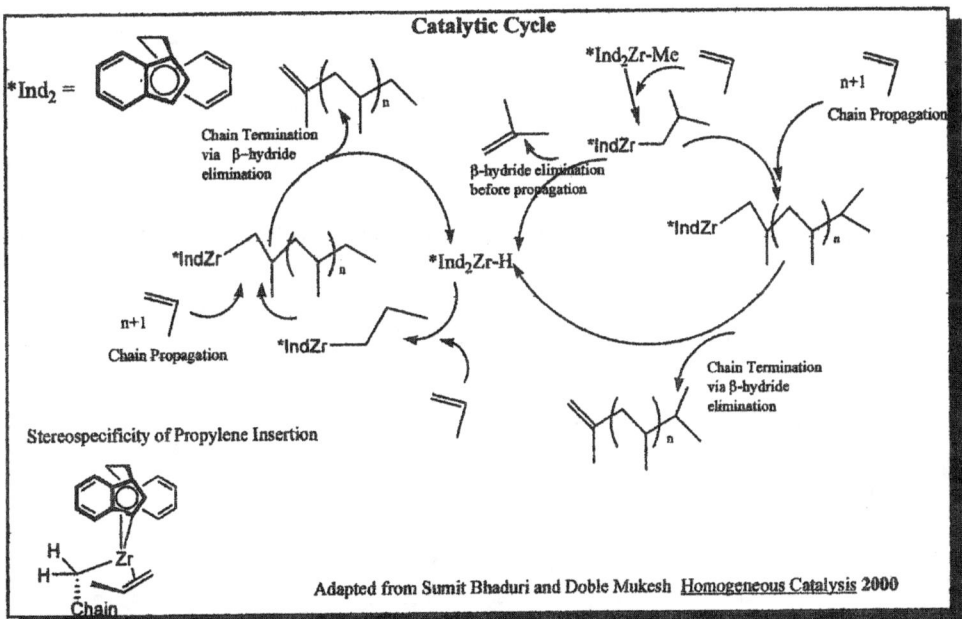

Figure 17 Green-Rooney Mechanism

confirmed (rather than merely proposed) for both the homogenous (soluble) and heterogeneous (insoluble) catalytic systems. Different interpretations of these details

40

have led to two main schools of thought on the exact mechanism of the more difficult

heterogeneous system, the Cossee-Arlman mechanism (direct insertion) and the Green-

Rooney mechanism (alkene to metal coordination), named after early pioneers of

mechanistic studies in this field [79].   However, for the simpler, homogeneous system,

the Green-Rooney mechanism (**Figure 17)** is generally accepted [79].

While a very nice collection of the original work in this field has recently been put

together by Blom and others [83], a much more succinct summary is found in the Bhaduri

and Mukesh textbook [79]. In point of fact, however, few differences between the

original proposed mechanisms and the confirmed mechanism exist. Even the advent of

MAO (methylaluminoxane) merely changes the nature of the aluminum and the bridging

groups while the addition of chirality makes the additions stereoselective. The metal

center is still the active site, the incoming alkene still enters the coordination sphere, is

still bound to the metal, and inserted between the metal center and the ethyl (or growing

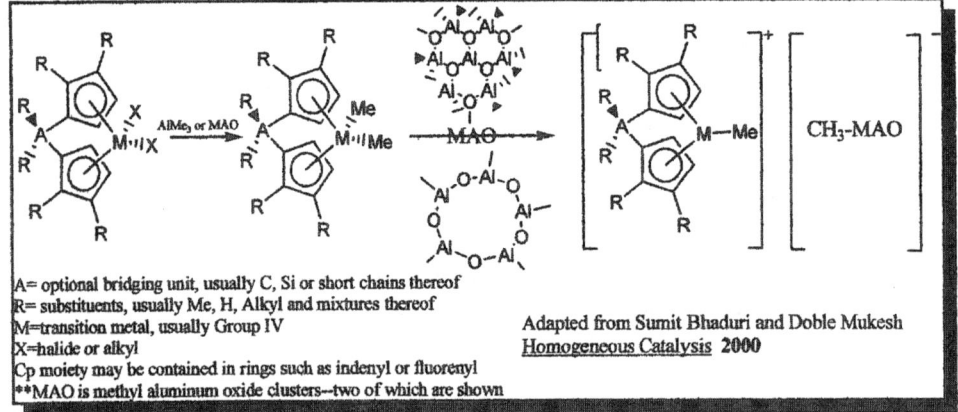

A= optional bridging unit, usually C, Si or short chains thereof
R= substituents, usually Me, H, Alkyl and mixtures thereof
M=transition metal, usually Group IV
X=halide or alkyl
Cp moiety may be contained in rings such as indenyl or fluorenyl
**MAO is methyl aluminum oxide clusters--two of which are shown

Adapted from Sumit Bhaduri and Doble Mukesh
Homogeneous Catalysis **2000**

Figure 18 Role of Methylaluminoxane

chain), and is still terminated by β-hydride elimination. The only real difference with

41

MAO is the nature of the co-catalyst and its effect on the active site by effectively isolating the methyl anion from the zirconocene cation [79]. This isolation allows the incoming olefin easier access and greater affinity to the binding site [79].

While there obviously has been a great deal of research done on the mechanisms of zirconocene interactions, much of the knowledge of the polymers came after the fact by reviewing the structure and function of the catalysts and the properties of the polymers produced [84]. For example, by examining polymers and the catalysts responsible for them, it could be generally summarized that C_2-symmetry in the catalyst is responsible for the production of isotactic polymers of propylene [95]. With such knowledge, the modern search for catalysts to form simple isotactic and atactic polymers is giving way to catalysts that are tailor-made for producing specific types of polymers, such as hemi-isotactic (wherein every other pendant group is on the same side of the chain and no set stereochemistry for the others) via C_1-symmetrical catalysts or syndiotactic polymers (wherein every pendant group is on the opposite side of the chain from its nearest neighbors) via Cs-symmetrical catalysts or even block polymers (wherein different stretches of the chain follow different types of tacticity in monomeric polymerization or even different pendant groups in co-polymerizations) [79, 165].

2.8 Conclusion

In conclusion, this chapter has shown that much work has been done on zirconocene catalysts since their inception. Development of catalytic systems and

42

knowledge as to the mechanisms and nature of polymerization results (tacticity,

molecular weight, etc.) has progressed dramatically since the early systems of Ziegler and

Natta. Better co-catalysts, such as MAO, have been discovered. Issues of chirality have

been added to both ligands and zirconocenes. Prochiral ligands, such as Brintzinger's

ethylene-bridged bis(indenyl) [EBI] and ethylene-bridged bis(tetrahydroindenyl) [EBTHI]

have been developed and the chiral metallocenes therefrom separated and applied to the

production of chiral oligiomers, as well as other asymmetric reactions. However, while

much has already been learned of the nature of catalysts and the polymers produced,

much work still remains to be done. The quest for a better catalyst is unceasing.

Chapter 3

Issues of Conformational Isomers

3.1   Introduction

As the previous chapter indicates, much has already been learned of the nature of catalysts and the types of polymers produced. However, while much is known, there are areas in which the records are lacking. One such area of research is in the discrete conformations of zirconocene catalysts. While it is known that the relative energies of a few conformations of zirconocenes play a role in the tacticity of polymers produced [25], the conformations of discovered catalysts still do not cover the full spectrum of defined conformations [1c]. This results in gaps of knowledge about the role of specific conformations in specific reactions.

While the effects of symmetry on the type of polymer produced was mentioned in the last chapter, other structural aspects have also been examined as to their effect on the activity of the catalyst. Modifications to the bridging unit and substitution patterns have all been studied [166] with differing degrees of success. Indirectly, these modification studies hint at the importance of learning more about conformations. As Kaminsky summarizes the result of some of these studies, "rising flexibility of the ligand framework induced by an enhanced polymerization temperature may destroy the preferred mode of catalyst geometry" [166]. In other words, as the temperature rises in the polymerization

44

process, the extra energy allows the catalyst to interconvert between conformations, some of which are less active and/or selective than others.

3.2   Torsional Isomers

While interconvertible isomers had been proposed in previous years to account for the variations of catalytic activity and selectivity at various temperatures [167,168, 169], Kaminsky and others applied a nomenclature system to conformational isomers [166]. Working with Brintzinger's EBI complex, Kaminsky realized that the five-membered ring composed of the bridging carbons, the 1 and 1' indenyl carbons, and the zirconium could form two distinct conformational isomers for each enantiomer. Using established ring

helicity assignments, he termed the two isomers δ (right-handed) and λ (left-handed) [166]. Since these two isomers are merely a ring twist apart, Kaminsky called them torsional (twist) isomers. The two (R,R)-EBI ZrCl$_2$ isomers are shown in **Figures 1** and **2** (adapted from Kaminsky's article [166]) as viewed down the zirconium. In this typical representation, the bonds of zirconium to

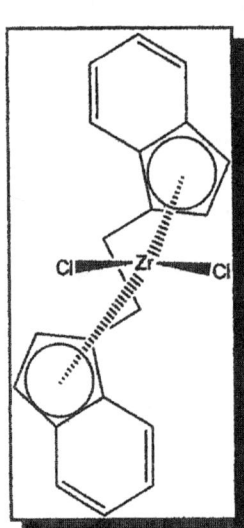

Figure 2 δ-(R,R)-EBI ZrCl$_2$

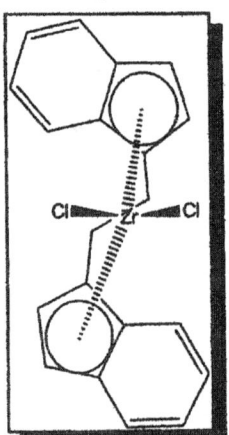

Figure 1 λ-(R,R)-EBI ZrCl$_2$

45

the five carbons of the cyclopentadienyl moiety are omitted for the sake of clarity, but are still existent. To clarify the matter of torsional isomers even further, imagine turning the structure until the chlorides are vertical and the bonds from the zirconium to the C1 and C1' on

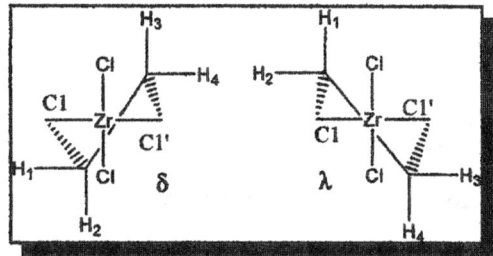

Figure 3 δ- and λ-Torsional Isomers

the indenyls are horizontal. Looking straight down the zirconium, the ethylene bond may be seen sloping to the right of the zirconium in one conformer and sloping to the left in the other. The conformer with the right slope is the δ(delta)-isomer and the one with the left slope is the λ(lambda)-isomer (**Figure 3**).

Perhaps the clearest view of the two conformers is the view looking down the top of the indenyls. The bond of the bridging carbon nearest the viewer then slopes to the right as it descends away from the viewer to the other carbon of the bridge for the δ-conformation and to the left of the viewer for the λ-conformation (**Figure 4**).

It is also worth noting that the

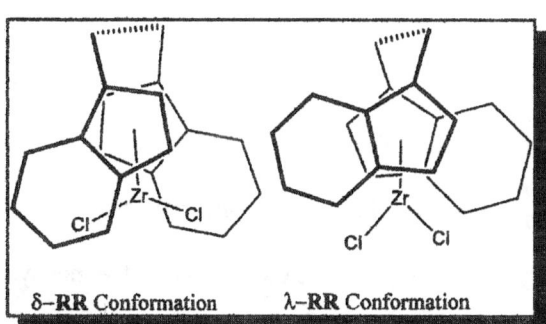

δ–RR Conformation   λ–RR Conformation

Figure 4 Top View of Torsional Isomers

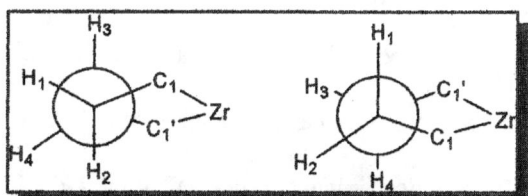

Figure 5 Newman projection of δ- (left structure) and λ- (right structure) Isomers

46

hydrogens of the ethylene invert their axial/equatorial orientation when undergoing the ring flip inversion. The axial/equatorial interconversion may be more easily seen in the more traditional Newman projection shown in **Figure 5**. Brintzinger and others had already established the rapid (NMR time scale) conversion of the two isomers in solution [167,168,170] which suggested the interconversion of two isomers. Of course, the polymerization mechanism involves a three-coordinate zirconium that is less sterically demanding than the four-coordinate metallocene dichloride and is usually performed at elevated temperatures. Interconversion, therefore, could occur with even greater facility under polymerization conditions [171].

So far, the indenyl orientation has been ignored. However, since the orientation of the indenyls is responsible for differing

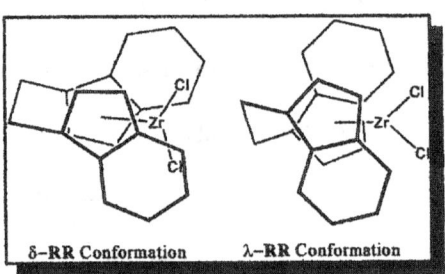

Figure 6 Torsional Isomers with Twisted Bridging

degrees of isotacticity and molecular weights in the polymer produced [166-170], it can not be ignored. **Figure 6** shows the indenyl orientations of the two *(R,R)*-conformers (double bonds are omitted for reasons of clarity). The difference in indenyl orientation is quite obvious. However, the effects are somewhat less evident. If the indenes are examined as the jaws of an alligator in two-dimensional space, the δ -conformer appears to be more closed and the λ-conformer more open. This phenomenon allows the zirconium to fit deeper into the "pocket" of the λ-conformer and should allow a more selective approach of incoming propylene units. Thus, the more "open" appearing conformation is

47

thought to be the more selective of the two conformers.

Kaminsky put this theory to the test by substituting methyls at the 2, 4, and 7- positions on the indenyl rings. In this manner, Kaminsky was able to achieve an analog to the Brintzinger complex that was only able to achieve the λ-isomer (**Figure 7**). This preference is assumed to be due to the greater steric

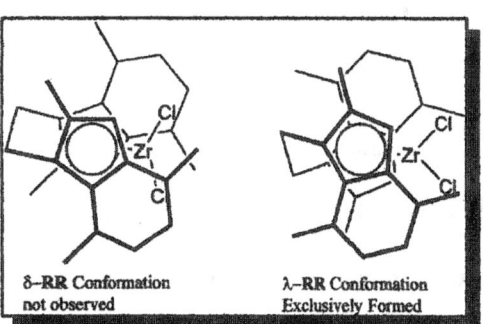

δ–RR Conformation
not observed

λ–RR Conformation
Exclusively Formed

Figure 7 Ethylene-bridged bis(2,4,7-trimethyl indenyl)zirconium dichloride

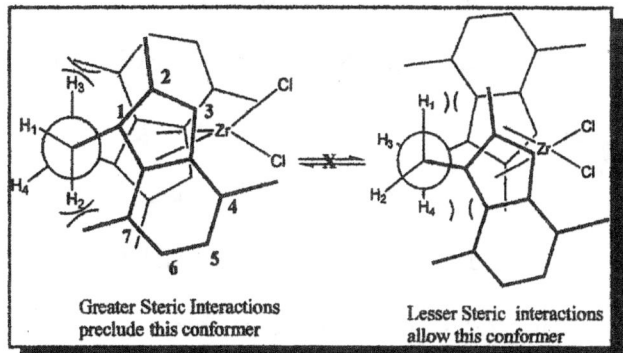

Greater Steric Interactions
preclude this conformer

Lesser Steric interactions
allow this conformer

Figure 8 Newman Projections of Kaminsky's Analog

interactions of the methyls at the 7-position than those of the 2-position with the hydrogens of the bridge, as shown in **Figure 8**.

Since isotacticity is dependent on the shape of the ligands controlling the incoming propylene [166], it would be reasonable that decreased flexibility would increase isotacticity. Similarly, since the β-hydride transfer that causes chain termination is more demanding, sterically, than olefin insertion [166], it would be reasonable that decreased flexibility would increase the molecular weight of polymers. And, indeed, Kaminsky's analog polymerized propylene with a much higher degree of isotacticity and to higher molecular weights than the fluctuating Brintzinger version [166].

48

3.3   Π and Υ Designations

While the twist nomenclature of δ and λ accurately distinguishes between the two

conformations of each enantiomer, it becomes somewhat less helpful when applied to the

actual structure of the molecules, especially when dealing with pairs of enantiomers and

the orientation of the indenyl rings. The confusion lies in the comparison of the

nomenclature and the absolute configuration

[95, 170] as λ-RR, is the mirror image of

the δ-SS and the δ-RR is the mirror image of

the λ-SS. To understand the dilemma,

reference back to the last section to find that

the λ-RR is the more "open" configuration

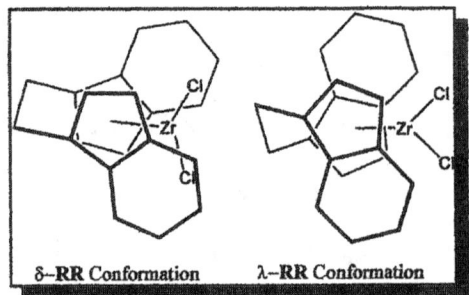

δ–RR Conformation   λ–RR Conformation

Figure 9 Closed/Open Conformations

and expected to be the more active conformer, but in the other enantiomer the δ-SS is the

"open" configuration. Again, since it is the actual structure to activity that is being

studied, it would be easier to reference the similarly structured conformers.

In order to accomplish this referencing Fabrizio Piemontesi, Angelo Sironi, et al

introduced positional nomenclature of Π (Pi) (προ, Greek for "in front") and Υ (Upsilon)

(υστερου, Greek for "behind") [170] to reference the position of the indenyl groups. This

system disregards the issue of torsional isomerism or enantiomerism and refers only to the

structural positioning of the indenyl as toward or away from the zirconium center. Of

course the torsional and positional naming systems can be correlated, as seen in **Figure 10**.

49

While research on the positional isomers of Brintzinger's complex has been published sporadically in the literature for nearly a decade [166-174 and many of the references therein], much remains to be learned. The most limiting factor on expanding this research is that in order to draw detailed information, complexes with specific, defined, rigid conformations need to be formed and examined in order to define the actual structure/activity relationships between the catalyst and the types of polymers produced.

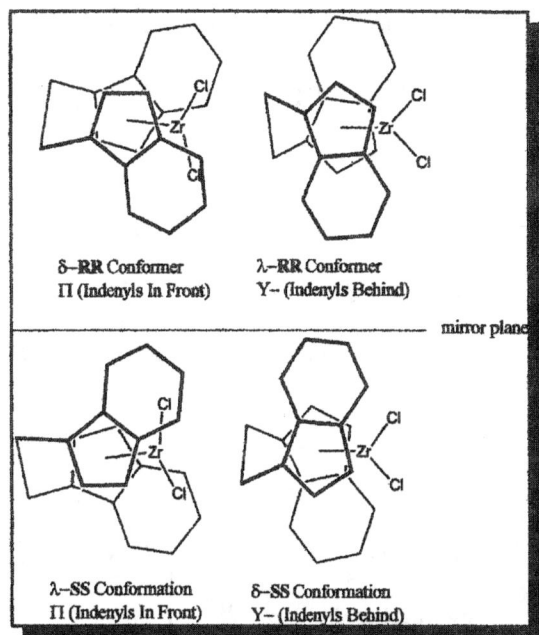

Figure 10 Torsional and Positional Conformers Correlated

3.4   Single Conformers

Kaminsky worked with tri-substituted indenyls, sterically locking out the Π- (δ-RR, λ-SS) conformations. This work resulted in the first successful attempt at

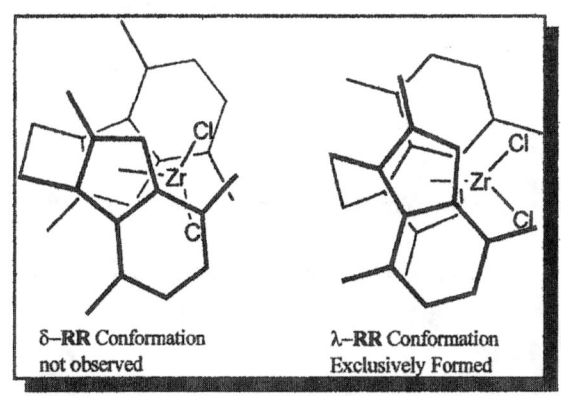

Figure 11 Kaminsky's Success

50

achieving a single, isolated conformer (**Figure 11**). However, while it was the first

successful attempt, it was not the first attempt.

One of the earliest attempts to form a single conformer is found in the work of

Bernhard Rieger [85]. Rieger approached the isolation of single isomers by considering

the bridging unit. If it could be modified in such a way as to preclude ring flips, then the

individual conformers could be isolated. To accomplish this feat, Rieger formed an analog

to Brintzinger's EBI
ligand with a trans-1,2-
cyclohexanediyl bridge
by forming ditosylate **25**
from diol **24** (**Figure 12**).
After the initial formation
of the axial ditosylate

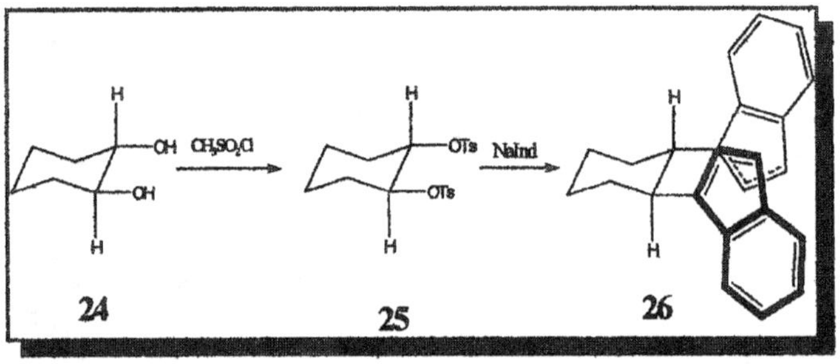

Figure 12 Rieger's trans-1,2-Cyclohexanediylbridged Ligand [85]

from the S_N2 displacement, the ring flips to the preferred equatorial position of ditosylate

25. With the cyclohexyl bridging unit, the ring flip required for the two indenyls to return

from equatorial to axial position is unlikely at any reasonable temperature in the ligand,

let alone the metallocene. Also, the indenyls, after SN2 displacement of the tosylates and

ring-flip back to the equatorial positions, are not able to rotate freely about the sigma

bonds connecting them to the ring system. Therefore, the stereochemistry that will result

from metallation is already set in the ligand. Upon actual metallation, a mixture of

51

diasterereomers and their enantiomers must form (**Figure 13**). Thus, the trans-(1R, 2R)-1,2-

bis(indenyl)cyclohexane ligand should give rise to not only inconvertible Π- and Υ-

diastereomers [(1-R, 2-R, 1-pS, 1'-pS) and (1-R, 2-R, 1-pR, 1'-pR), respectively], but also

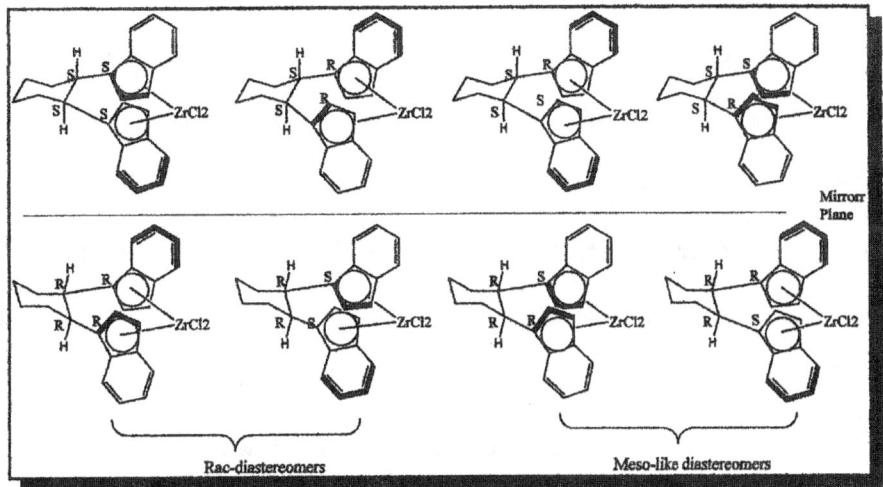

Figure 13 Rieger's Four Diastereomers and their Mirror Images

two other diastereomers, [(1-R, 2-R, 1-pS, 1'-pR) and (1-R, 2-R, 1-pR, 1'-pS)] and the

trans-(1S, 2S) ligand can give a similar set, as seen in **Figure 13** (shown in the less usual

line-eclipsing format for easier determination of planar chirality assignment). The more

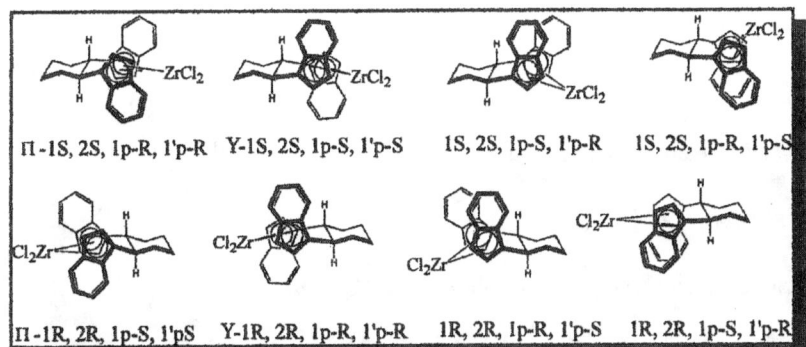

Figure 14 Standard View of Rieger Mixture

52

standard view (**Figure 14**) shows the positional/conformational nature of the various diastereomers more clearly. It might be noted that the apparent *meso*-like isomer does not have a mirror plane of symmetry if the cyclohexane ring is taken into account. Therefore, like *meso*-(EBI)ZrCl$_2$, it is in a "chiral conformation," [170] but unlike *meso*-(EBI)ZrCL$_2$, it is not able to interconvert. Thus, all four isomers and their enantiomers may be formed and in fact are [85]. Unfortunately, the mixture of isomers was inseparable. However, Rieger attempted polymerization with the mixed isomers and, like Kaminsky, found higher polymerization activity in the non-flexible catalytic system, even with the large mixture of trans-1,2-cyclohexyl-bridged bis(indenyl) zirconium dichlorides, than with Brintzinger's (EBI)-zirconium dichloride [85]. Since one isomer is usually more active than the others, this discovery of high activity in the mixture suggests fairly strongly that one of the isomers is much more active than the others: the average isotacticity and molecular weight being diluted by the effects of the less active isomers in the batch. If the single isomers could have been separated, much information could have been gleaned.

Besides steric modifications to the indenyl {such as Kaminsky's trimethyl-substituted ansa-bis(indenyl) **27** [166]}, and modifications to the bridge {such as Rieger's cyclohexyl- bridged ansa-bis(indenyl) **28** [85], Brintzinger's etheno-bridged

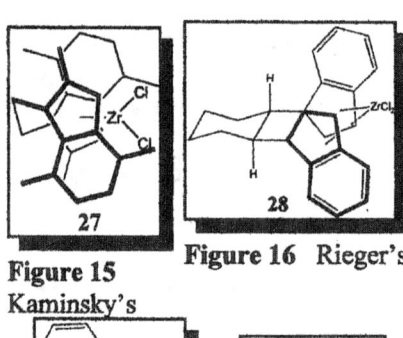

Figure 15
Kaminsky's

Figure 16 Rieger's

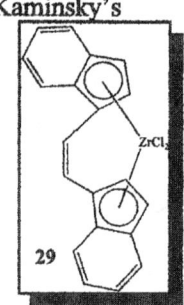
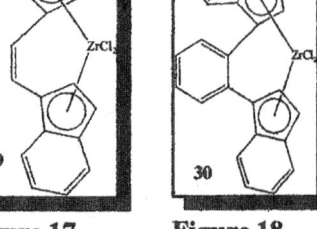

Figure 17
Brintzinger's

Figure 18
Halterman's

53

ansa-bis(indenyl) **29** [173], and Halterman's

phenyl-bridged ansa-bis(indenyl) **30** [95]},

attempts to render the ansa-bis(indenyl)

metallocenes more rigid and more likely to exist in

a single conformation have most often involved

shortening the bridging unit to a single unit, often

methyl or silicon [95] (**Figure 19**). However,

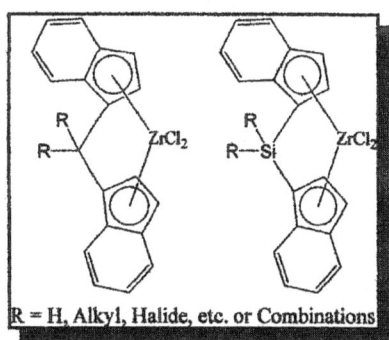

Figure 19 Most Common Method of Introducing Rigidity

other methods, including increasing the number of atoms in the chain of the bridging unit

[172], using double-bridges [95,175] and changing the position of the bridging units [176],

have all been used with varying degrees of success.

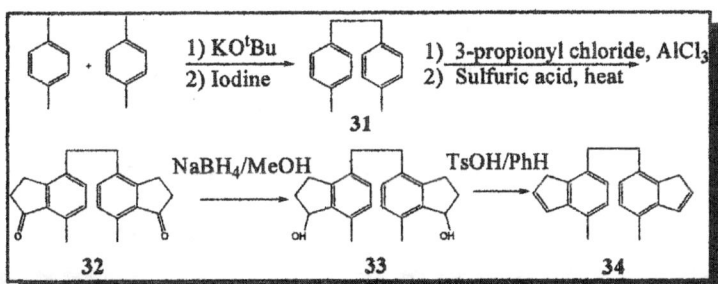

Figure 20 Novel Route to a Ligand Giving Very Open Metallocene

Two of the routes to successful single conformational

isomers are especially worth mentioning. A route by

Halterman and Combs to a very open metallocene involved a

novel double Friedel-Crafts combined acylation/alkylation

reaction on **31** [87, 95] to generate **32** and, eventually, a C7-

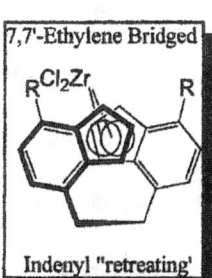

Figure 21 Very Open Metallocene

54

C7'-ethylene bridged bis-indenyl ligand (**34**) that metallates to a single conformational

isomer of a much more "open" nature than in typical metallocenes (**Figure 21**) [176].

The other important route to a single conformational

isomer, by Halterman, Tretyakov and others, metalates to a

much more closed position (**Figure 22**) than seen in typical

metallocenes, such as the Brintzinger (EBTHI)ZrCl$_2$ [175].

This metallocene was generated via an intriguing route

involving oxidation of 2,6-dimethyl-1,5-cyclooctadiene (**35**) to

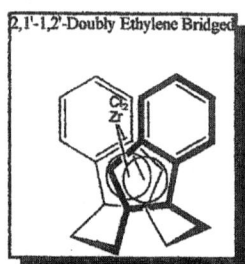

Figure 22 Very
Closed Metallocene

36, followed by a double Grignard addition to the diol and Swern oxidation back to the

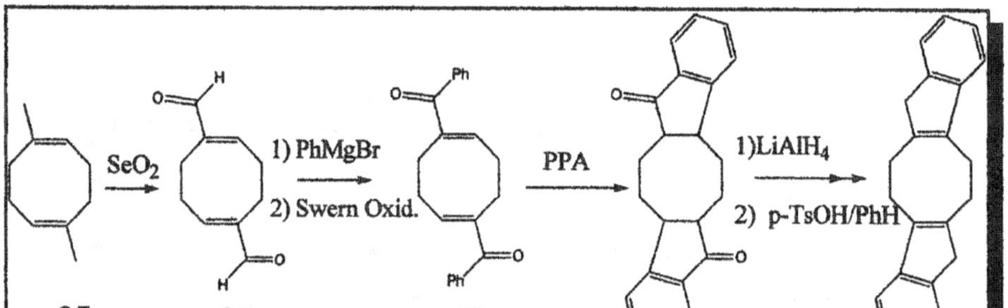

Figure 23 Novel Route to a Ligand Giving Very Closed Metallocene

conjugated ketone **37** which could be closed with a double Nazarov closure, yielding (after

reduction and elimination) a doubly-bridged ligand (**39**) [88, 96] that could be metallated

to the very closed position seen in **Figure 22.**

The importance of these two single conformational metallocenes lies in their

extreme nature. The 7,7'-bridged metallocene is even more "back" than the Brintzinger ϒ-

conformer, with the indenyls actually retreating away from the metal center (here dubbed

55

E, epsilon, for the Greek εξίδος or retreat), while the doubly ethane bridged 1,2' and 2,1' is even more closed than the Brintzinger II-conformer, seeming almost to clamp the metal

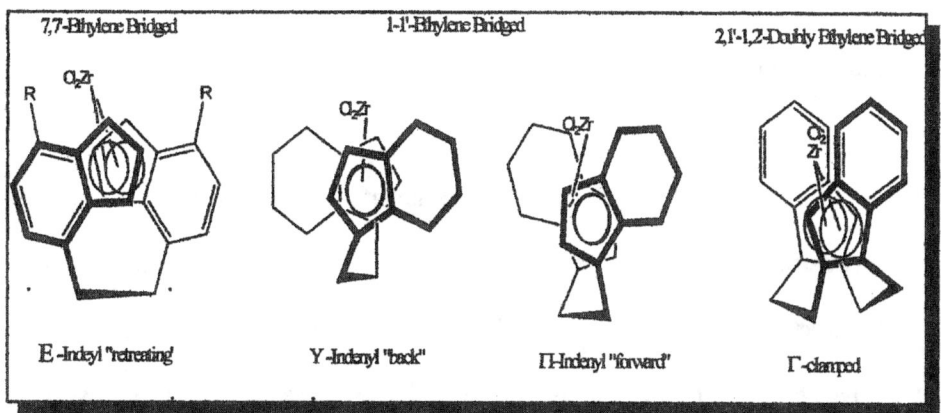

Figure 24 Conformational Representations

center (here dubbed Γ, gamma, for the Greek γόμφος or clamp). These single conformational isomers mark the limits of the known geometries. Unfortunately, to date, the polymerization data has not been established, so the actual structure to activity relationships of these single conformational metallocenes remains unknown.

As a final note, with the advent of bridges in positions other than the typical Brintzinger position, the bridging position can no longer always be assumed to be the 1-position. IUPAC has therefore introduced a naming system in which the indenyl rings are numbered as indenes and the position at which the bridging unit is attached inserted between the inden- and the-yl. For example, the typical Brintzinger ligands would be 1,2-bis(inden-1-yl)ethane and 1,2-bis(tetrahydroinden-1-yl)ethane. Halterman's 7,7' ethylene bridged ligand would be 1,2-bis(inden-7-yl)ethane. This IUPAC-suggested naming system is still not universally applied and many other naming systems can be found in the

56

literature. However, the IUPAC system provides a comprehensible format when other naming systems are not already in use.

3.5   Conclusion

In previous chapters it has been seen that the uses of ansa-bis(indenyl) metallocenes are multitudinous, ranging from asymmetric synthesis to carbon-fluorine bond activation and cyclizations. The historical development of the metallocenes has been driven primarily by their use as polymerization catalysts and shows the vast array of research that has been done in developing zirconocene catalysts. Issues of chirality have heightened interest in the chiral metallocene field for both enantioselective reactions and polymerizations. In this chapter, structure to activity relationships was introduced. The existence of fluctuation in the classic Brintzinger EBI and EBITH zirconocene complexes introduced conformational isomers. The increased reactivity, stereoselectivity, and molecular weights achieved by Kaminsky's single conformational isomer and the heightened reactivity of Rieger's mixture introduced the need for more knowledge about the structure/activity relationships of specific conformers. Halterman's research has established the known limits of defined single conformational isomers. However, gaps still remain in the intermediate ranges.

57

Chapter 4

Conformational Contortions of Unusual Bridging Positions

4.1   Introduction

As previously mentioned, Halterman's known, extreme, rigid conformations, set

the limits of conformation with retreating (**Compound 40**) and clamped (**Compound 43**)

metallocene structures [87, 88, 95, 96]. Brintzinger's *ansa*-bis(inden-1-yl) ethane has

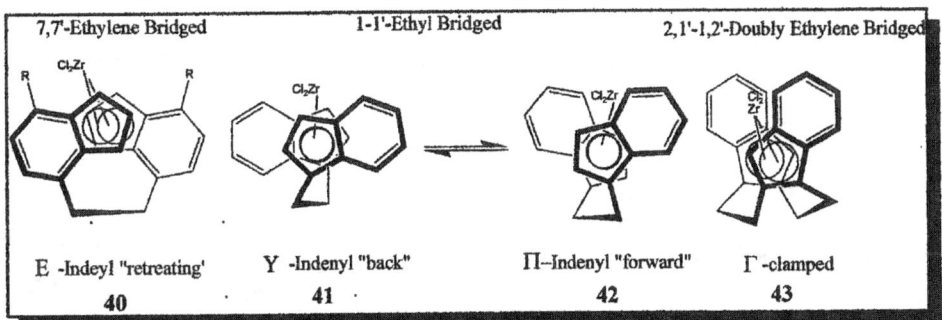

Figure 1 Known Conformations

intermediate conformations, but the conformations are in constant flux [166] between the

"indenyl-back" conformer **41** and "indenyl-forward" conformer **42**, and therefore do not

"provide a well-defined geometry" [175] for structure to activity studies. The Rieger

mixture should have rigid, defined intermediate conformations that should not

interconvert at any reasonable polymerization temperature, and it was quite active

towards polymerization. However, the diastereomeric mixture of metallocenes was

58

inseparable [85] (**Figure 2**) and the more active structure impossible to absolutely

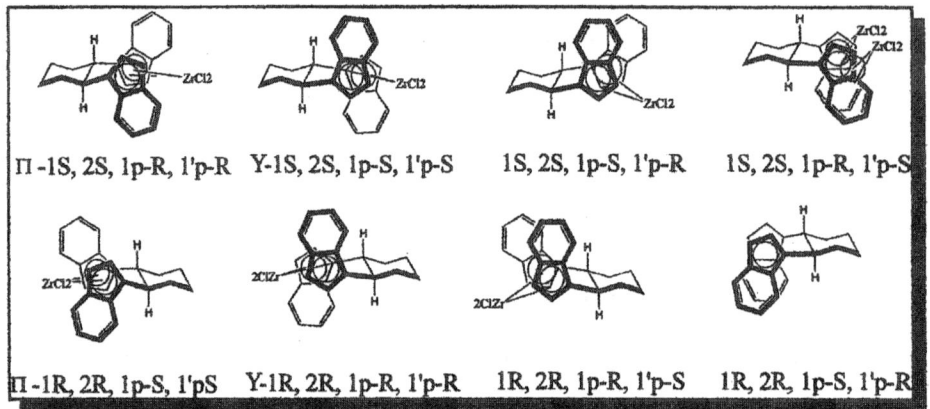

Π-1S, 2S, 1p-R, 1'p-R   Y-1S, 2S, 1p-S, 1'p-S   1S, 2S, 1p-S, 1'p-R   1S, 2S, 1p-R, 1'p-S

Π-1R, 2R, 1p-S, 1'pS   Y-1R, 2R, 1p-R, 1'p-R   1R, 2R, 1p-R, 1'p-S   1R, 2R, 1p-S, 1'p-R

Figure 2   Rieger's Active but Inseparable Mixture

determine. Kaminsky's 2, 4, 7-trimethyl-substituted bis(indenyl) zirconocene dichloride

hints tantalizingly that the λ-**RR**/δ-**SS** structures (Υ-conformations) are more active and

selective. However, it is generally
known that 2-substituted bis(indenyls)
are more active and selective than the
non-substituted analogs [166], so his
results are inconclusive without either
the results for the Π-conformations or the

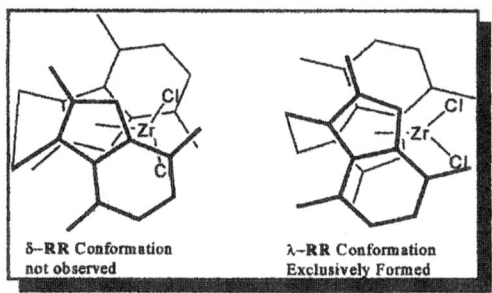

Figure 3   Kaminsky's Active Complex

analog not substituted in the 2-position, as the increased activity could come from the

substitution at the 2-position.

In order to advance the knowledge of structure/activity relationships, we decided

to examine bis(indenyl) zirconocenes of rigid, intermediate conformation. The design of

these compounds considered the bridge-linkage sites, chain length, and steric interactions.

59

As the basis for these studies, we undertook a project involving an inden-7-yl

species with a side chain that

could be modified to form a two

or three carbon bridge from the

7- to 1'- positions in bridged

bis(indenyl) metallocenes. Such

metallocenes are expected to

evince rigid conformations such

that a single conformational isomer

(Y-conformers **44** and **46**) should

be formed with a structure

intermediary between the known

extremes of the fully closed

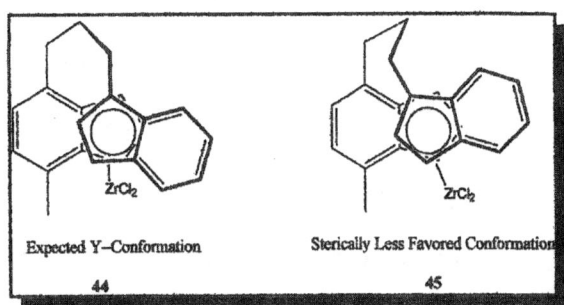

Expected Y–Conformation   Sterically Less Favored Conformation

44   45

Figure 4 7,1'- Propylene-Bridged Complex

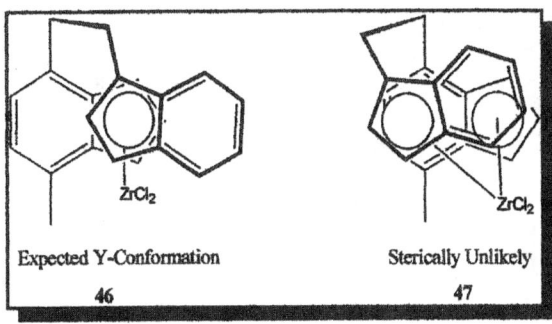

Expected Y-Conformation   Sterically Unlikely

46   47

Figure 5 7,1'-Ethylene-Bridged Complex

binaphthyl-bridged-2-2'-bis-indenyl zirconocene **43** [87, 95] and the fully open 4,4'-

diisopropyl-7-7'-bis-indenyl zirconocene **40** [88, 96]. The meso-like structures are

expected to be significantly higher in energy and should not be a factor.

Essentially, these rigid, intermediate conformations should fill in gaps in the

literature as to structure, and, when subjected to polymerization experiments, could yield

much information as to structure/activity relationships. The rationale behind these

expected conformations lies in the steric hindrance of the 4-methyl group to discourage

the meso-like isomers and the constraints of the bridging position to discourage the

formation of the torsional/conformational isomers. Additionally, the propylene-bridge in

60

the less likely E-conformation of the propylene-bridged complex would interact with the hydrogens of the 7-position of the inden-1-yl ring and the 1-position of the inden-7-yl ring (**Figure 6**), while the hydrogen at the 2-position of the inden-1-yl ring in the ethylene-bridged complex would interact with the pi system of the inden-7-yl ring in the less favorable conformation of the ethylene-bridged complex. In both cases (Complexes **44** and **46**), the expected conformation is near the Y-conformation that Kaminsky found to be so active [166], but without the methyl in the 2-position that confused his results.

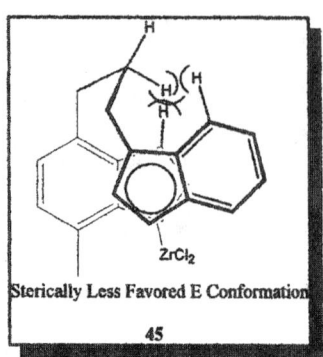

Figure 6 Hydrogen Interactions of Propylene-bridged Conformer E

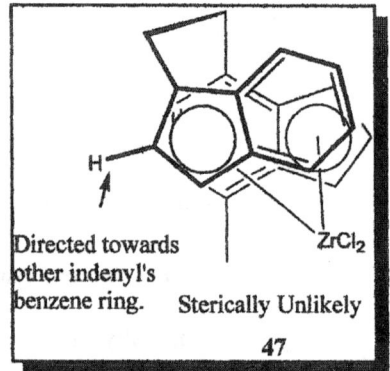

Figure 7 Hydrogen Interaction of Ethylene-bridged Conformer E

In addition, in the 7-2' ethylene-bridged system both torsional isomers should exhibit a Π-conformation (**Figure 8**). Like the meso-like Brintzinger Complex, this complex should rapidly interconvert conformations on the NMR timescale (though possibly slower as the zirconacycle forms a six-membered ring here rather than the usual five-membered ring). With the rapid

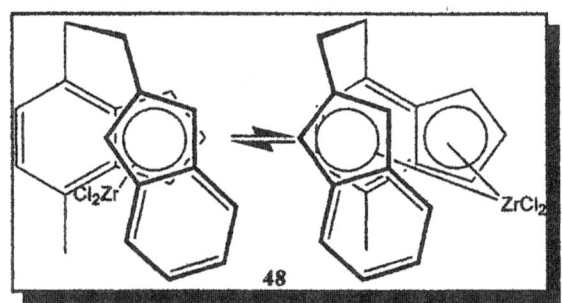

Figure 8 Torsional Isomers of 7,2'-Metallocene Complex

61

interconversion of these two conformations, metallocene **48** should essentially act as a single isomer. Also, due to the symmetry of the inden-2-yl ring, the possibility of meso-like isomers does not even exist. Of course, the substitution in the 2-position, in this case the bridging unit itself, does confuse the issue of activity. If the complex were to show exaggerated activity, the 2-position could be the cause and the issue unresolved. However, if the activity of this complex is lower than typical Brintzinger systems even with the 2-position substitution, then this conformation will be known to be of lesser activity. Together, the 7-1' ethylene- or propylene-bridged bis-indenyl zirconium dichloride with the 7-2' ethylene-bridged bis-indenyl zirconium dichloride system could solve, or at least add to the knowledge of, the issue of structure/activity relationships. We therefore undertook the synthesis of such compounds.

4.2   Cinnamates

In order to form the first of these new ligands that should yield single conformational isomers upon metalation, a route was examined that had the potential for chiral bridges as well. This initial pathway (**Figure 9**) was to take advantage of an aldol condensation between ethyl pyruvate and benzaldehyde to compound **49**. A Michael addition of an indenyl anion to **49** should form compound **50**. Saponification of **50** to compound **51** was to be followed by a Stetter Reaction [98, 99] with methyl vinyl ketone to compound form **52**. Erker's Indenyl Cyclization (1,4-diketone to indenyl by double addition) [100] should convert **52** to the desired ligand **53**. Metalation of **53** would yield

62

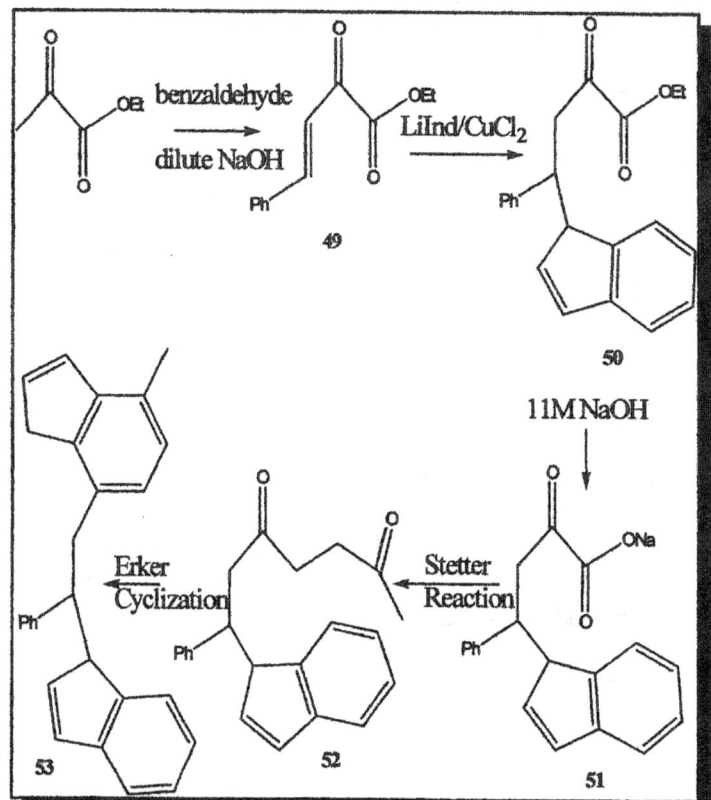

Figure 9 Proposed Cinnamate Route

a metallocene comprised of bis(indenyl)ligands with an ethylene bridge from the 7-position of one indene to the 1'-position of the other with a conformation very near the Y-conformation (**Figure 10**).

The advantages of this pathway include cheap starting materials, short sequence to ligand, novel methodology to a uniquely tethered bridged ligand, and the potential of achieving a single enantiomer via

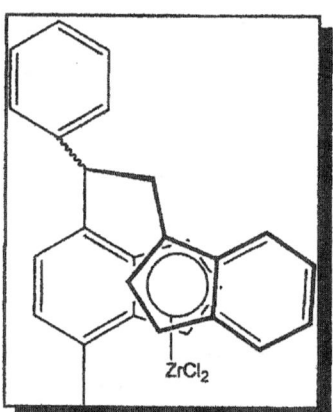

Figure 10 Cinnamate-based Metallocene

63

Corey's enantiospecific Michael addition [126] which could set the stereochemistry of the phenyl in the bridging unit.

This first proposed pathway underwent almost immediate modification. The aldol condensation under optimized conditions (under nitrogen in sodium hydroxide saturated ethanol) gave ~80% of what should have been compound 49. However, the ^{1}H NMR spectrum corresponded to the known

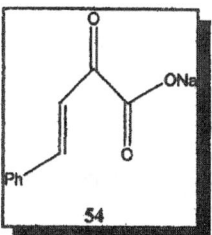

Figure 11 Known Salt

sodium salt 54, rather than the expected ester. This one-step saponification/aldol condensation seemed like a good thing, possibly eliminating the later saponification step. Unfortunately, the Michael Reaction with indene and potassium tert-butoxide on the sodium salt 54 failed to achieve any appreciable amount of product. An alternative route was needed.

The next alternative route that we examined focused on trans-cinnamaldehyde. In

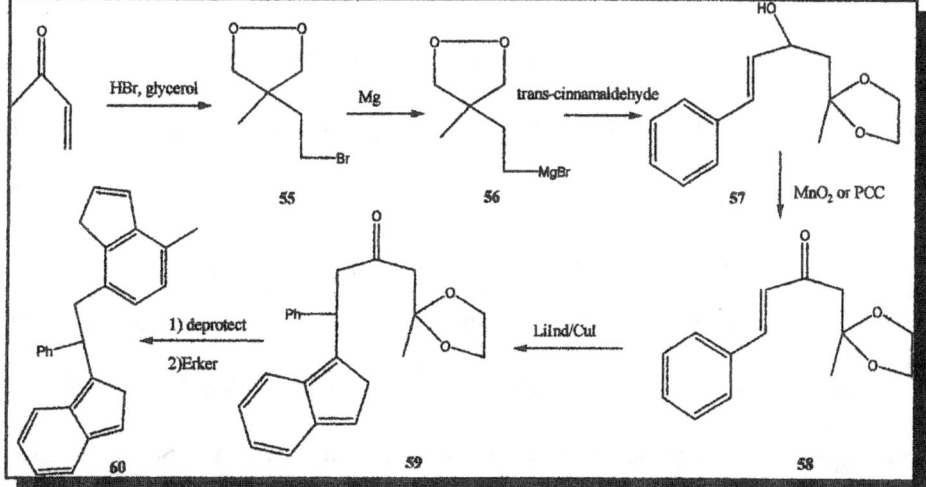

Figure 12 trans-Cinnamaldehyde route to 7,1'- ethylene-bridged bis(indenyl) ligand

64

retrospect, perhaps we should have performed the Stetter Reaction [98, 99] on trans-cinnamaldehyde to the 1,4-diketone, but instead we formed Grignard **56** [129] and added it to trans-cinnamaldehyde, as shown in **Figure 12**.

The first step in the above synthesis is the formation of ketal **55**, which proceeded according to the literature [128] in 90% yield. As seen in **Figure 13**, the ^{1}H NMR spectrum of the crude compound indicates that the compound is reasonably pure. In the literature, only the integration and chemical shifts are given (from a 25 MHz NMR), but the splitting patterns are as expected. The expected 3H (three hydrogen) singlet

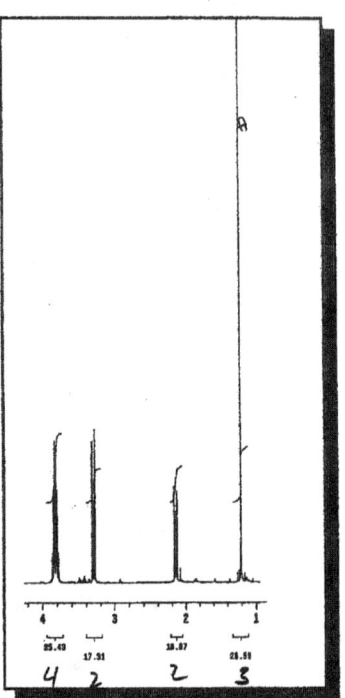

Figure 13 Crude ^{1}H NMR of Compound **55**

from the methyl group shows up as a 3H singlet at 1.25δ. The expected 2H triplet from the methylene beta to the bromide and ketal shows up as a 2H triplet at 2.15δ. The expected 2H triplet from the brominated methylene shows up as a 2H triplet at 3.30δ. The four hydrogens of the ketal are expected to be a 4H multiplet (four ddd at nearly the same shift), and we found a 4H multiplet at 3.85δ.

The Grignard formation from the product ketal was also known in literature [129], and, though touchy, proceeded in essentially quantitative yield. This yield was determined by quenching an aliquot of the solution containing **56** and comparing the H[1] NMR spectrum of the quenched compound with the spectrum of compound **55**. The

65

complete absence of the peaks from the hydrogens on the brominated carbon was taken as proof of the reaction's completion.

A large excess of the Grignard was cannulated to neat trans-cinnamaldehyde. The reaction was followed by TLC until no aldehyde was visible. An aliquot taken for ^{1}H NMR analysis proved the aldehyde had been mostly consumed in three hours. The aliquot was purified by column chromatography to separate product and remaining

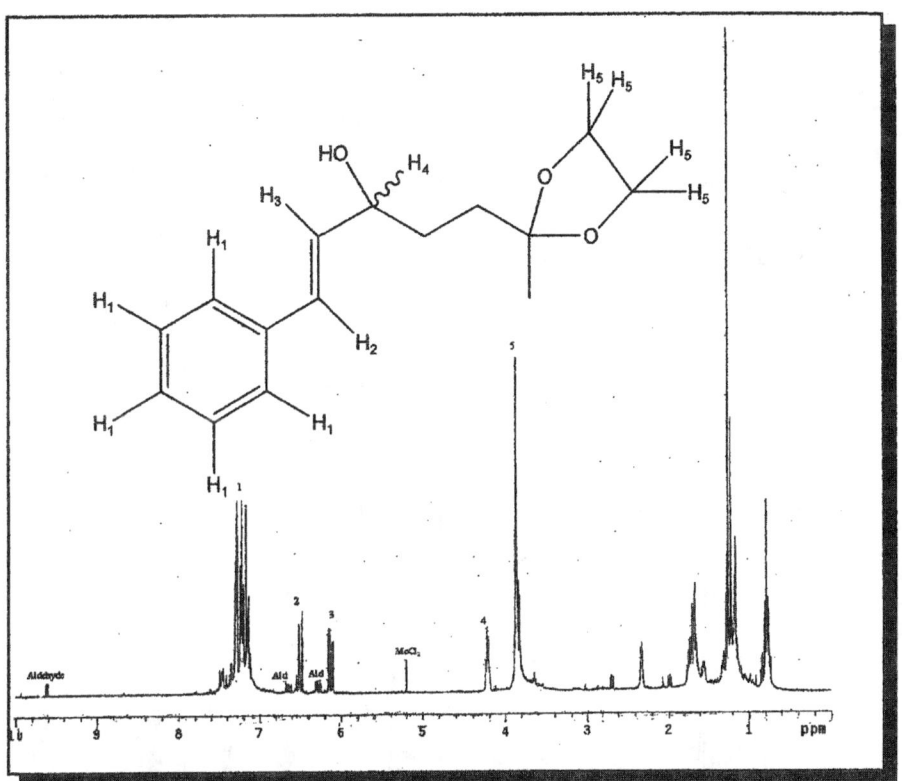

Figure 14 trans-Cinnamaldehyde and alcohol **57**

aldehyde (**Figure 14**) from the quenched Grignard remnants and other impurities. The conjugated alkenyl hydrogens that appear in the 6-7δ region are of particular significance,

66

exhibiting the expected (**Figure 15**) doublet and doublet of doublet at a slightly lower shift value (**Figure 16**) for the alcohol than the trans-cinnamaldehyde with similar coupling constants ($J_{1-2} = 15$, $J_{1-3} = 8$, $J_{2-3} = < 0.5$ – appears as a broad doublet unless greatly expanded).

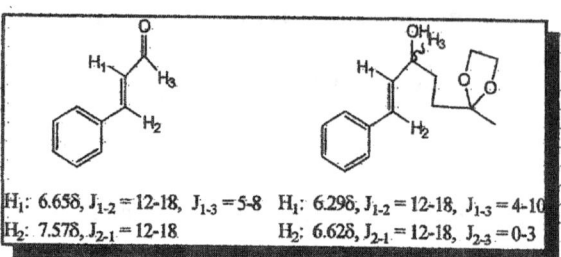

H_1: 6.656, $J_{1-2} = 12-18$, $J_{1-3} = 5-8$ H_1: 6.296, $J_{1-2} = 12-18$, $J_{1-3} = 4-10$
H_2: 7.576, $J_{2-1} = 12-18$ H_2: 6.626, $J_{2-1} = 12-18$, $J_{2-3} = 0-3$

Figure 15 Predictions from Silverstein Text [176]

With the inability to separate the alcohol completely from the aldehyde, the crude mixture, still under argon, was submitted to various oxidation

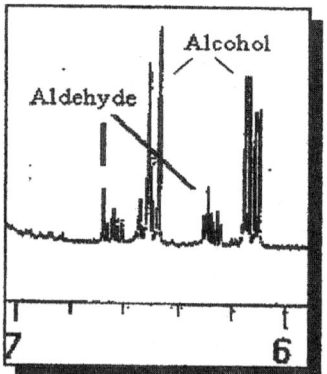

Figure 16 Alcohol 57's Alkenyl Peaks

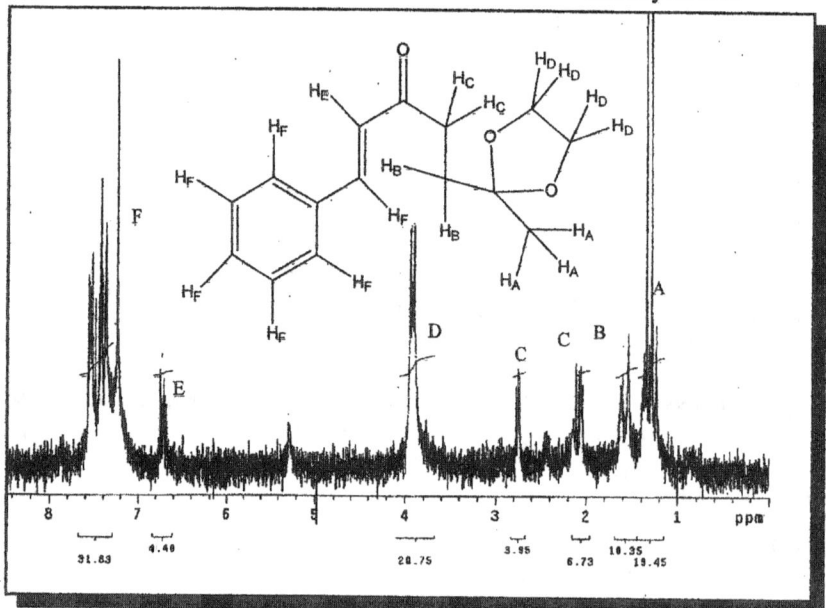

Figure 17 Chemical Shifts and Integrations of **58**

67

techniques to effect the transformation of the allylic alcohol to conjugated ketone **58**. After attempting several routes specific to allylic alcohols, the best oxidation technique used was simply the standard PCC (pyridinium chlorochromate) oxidation in methylene chloride. It is worth noting, that in the ^{1}H NMR spectrum (**Figure 17**), the ketone's

alkenyl hydrogens (**Figure 18**) show up very similar to the Silverstein-predicted values (**Figure 15**) for the trans-cinnamaldehyde, 6.7 δ and 7.4 δ versus 6.6 δ and 7.6 δ. Chemical shifts and integrations also align as might be expected (**Figure 17**). However, the yield of purified ketone (after purification via silica gel column chromatography) was a miserable 4% yield from the trans-cinnamaldehyde. After two runs through the synthesis, 0.2205 grams of purified ketone **58** was stockpiled.

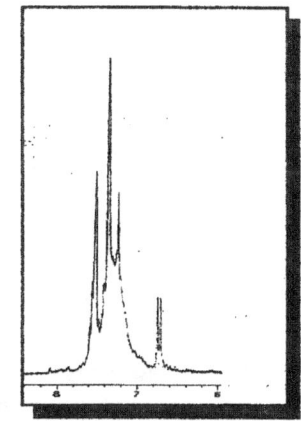

Figure 18 ^{1}H NMR of Ketone **58**'s Alkenyl Region

Since the product was so precious, a model reaction was examined for the Erker cyclization [100]. Since indenes are less stable, the indenyl anion Michael addition was to wait until after the Erker cyclization [100] was optimized with the model reaction. However, the model reaction led to a successful ligand development in its own right and this low yielding route became superfluous. A single attempt at the Michael addition followed by the Erker cyclization [100] with the entire stockpile gave no clear results and this rather disappointing pathway was dropped.

68

4.3   Propylene-bridged 1-(7-Methylinden-4-yl), 3-(inden-1-yl)zirconium(IV) dichloride

Combining the Stetter reaction [98, 99] considered earlier with the model study needed for the Erker cyclization [100], would form a precursor to other bis-indenyl

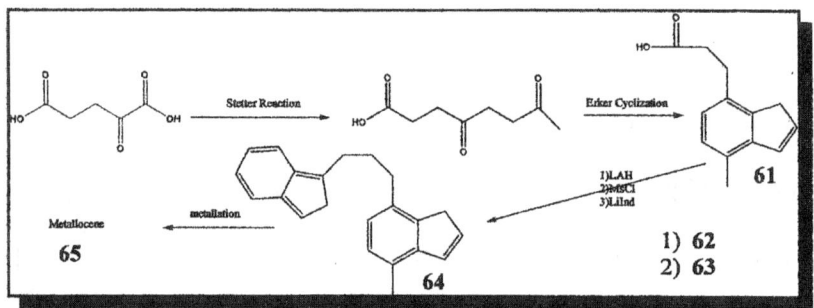

Figure 19 Route to Propylene-bridged Metallocene **65**

bridged ligands comprising a new route to potential ligands. While this new route (**Figure 19**) did not include the potential for chiral bridging units, it nevertheless had the

other values associated with the ethyl pyruvate and cinnamaldehyde routes and should still form a ligand likely to metalate to a single conformational isomer (ϒ-conformer) (**Figures 20** and **21**).

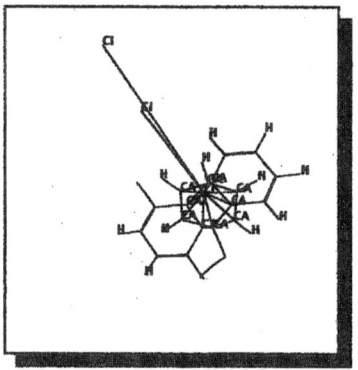

Figure 21 PC Model Minimum Energy for 1,3- Propylene-bridged bis(Indenyl) Zirconium Dichloride **65**

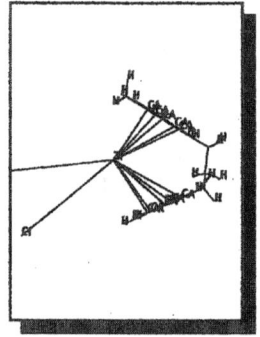

Figure 20 "Pocket" View of Complex **65**

69

This pathway has at its heart 3-(inden-1-yl)-propanoic acid (Compound **61**). From

this product the

propylene bridge could

be formed as shown

above. This acid

could be also be

esterified to **66** and

reacted with a di-

Grignard [101] to form

alcohol **67** and, after

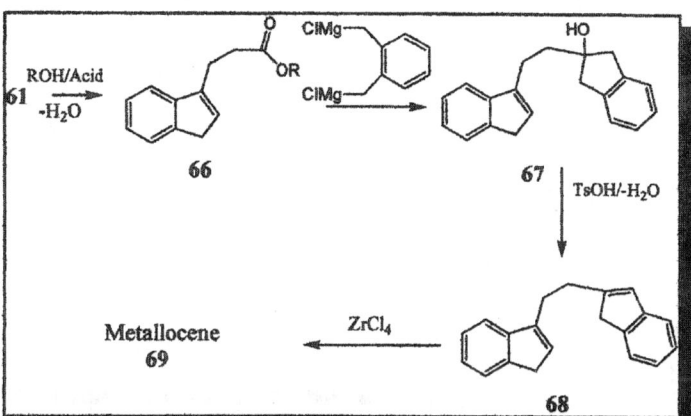

Figure 22 Route to Ethylene-bridged 7,2'-bis(indenyl)ZrCl$_2$ to be visited in Section 4.4

loss of water, the resulting ethylene-bridged 7-2' bis-indenyl ligand **68** (**Figure 22**) (and

revisited in detail in Section 4.4).

Indeed, the quick success and versatility of this route immediately changed it from

the originally proposed model study to the predominate pathway under consideration.

The high yields and facile reactions held great appeal and followed almost exclusively the

originally planned route.

The synthesis of the 4,7-dioxooctanoic acid (**Figure 23**), for example, proceeded

according to the literature

method of Stetter and

Lorenz [98] using

dioxane (or ethanol or

methanol with lower

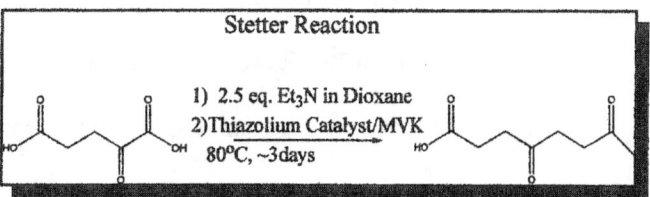

Figure 23 Conditions of the Stetter Reaction

70

yields) as solvent
and 10 mole % of
a thiazolium
catalyst (**Figure
24**) to
decarboxylate the
oxoacid end of 2-
keto glutaric acid
and add it to

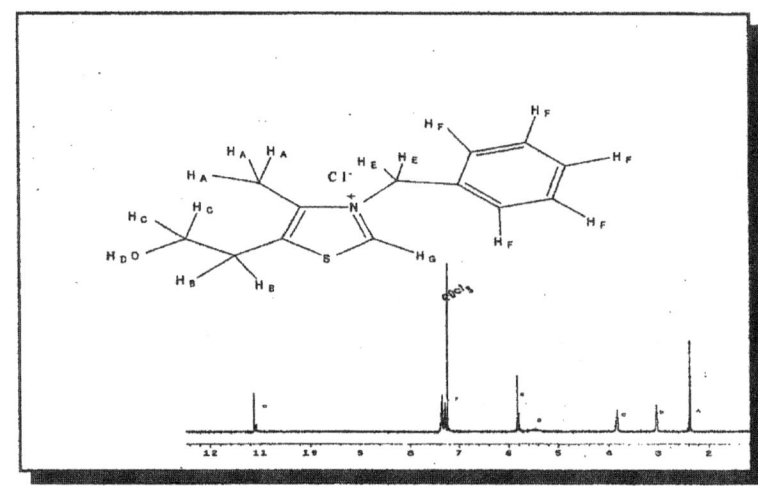

Figure 24 Thiazolium Catalyst's ^{1}H NMR Spec.

methyl vinyl ketone (MVK) at elevated temperature.

This route was the primary method by which this compound was generated.

However, later in the
synthesis a
modification of this
procedure (**Figure 25**)
by Novák et al[99],

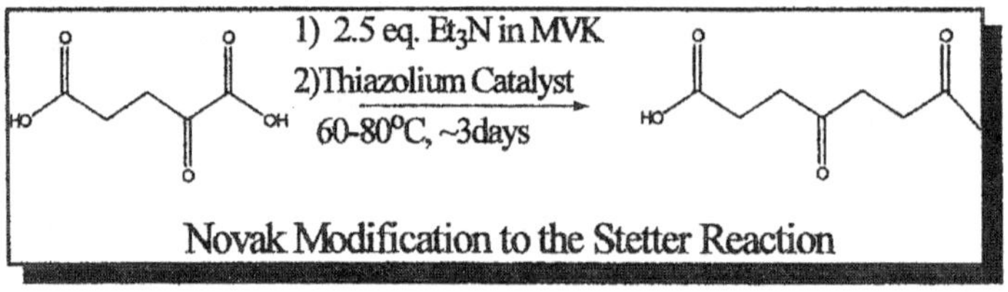

Figure 25 Modified Stetter Reaction

using methyl vinyl ketone itself as the solvent, was discovered in searching the literature.
This second route gave much higher yields. Common yields for Stetter and Lorenz's
method ranged from 17-35% on 5-20 gram scales, 35-42% on 1-5 gram scales, and up to
55% on scales of less than a gram. Using Novák's method, isolated yields of 55-60%
were common even on larger scales.

The reason for the dramatic differences in yields is assumed to be due to loss in

71

the aqueous work up. The presence of three carbonyls (one acid and two ketones) makes the compound quite water soluble, even in strongly acidic solutions. Thus, in the Novák method where methyl vinyl ketone (MVK) is the solvent, excess "solvent" is removed by simply placing the reaction mix under high vacuum, then washing with much smaller quantities of aqueous acid in the work-up/acidification of the residue.

After forming the known 4,7-dioxooctanoic acid and confirming its identity via ^{1}H NMR spectrum (**Figure 26**), Erker's route to substituted indenes (**Figure 27**) [100] was used to convert the diketone to the indenyl-substituted propanoic acid **61** and its double bond isomer (**61a** and **61b**).

The double-bond isomers, 3-(7-methylinden-4-yl)-propanoic acid **61b** and 3-(4-methylinden-7-yl)-propanoic acid **61a**, are formed in ratios ranging from 45:55% to 60:40%. The ratios of the two

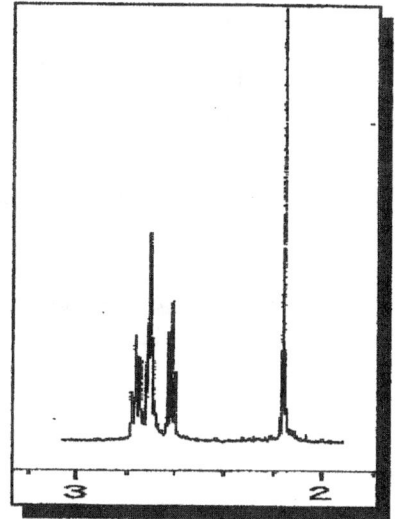

Figure 26 ^{1}H NMR of 4,7-dioxooctanoic acid

CpH/NaOMe

methanol

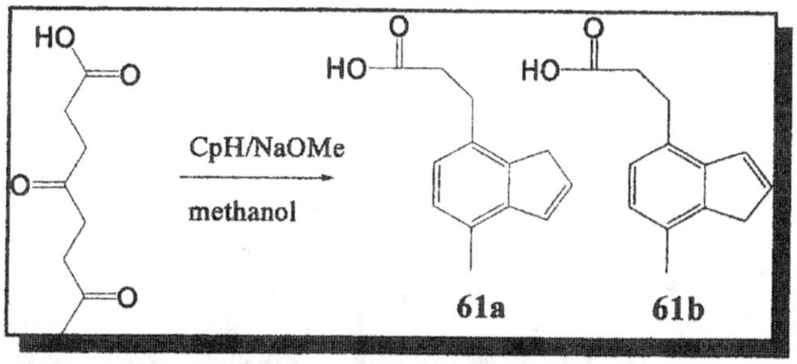

61a   **61b**

Figure 27 Erker's Cyclization of 1,4-Diketones to Substituted Indenes (Erker Indenyl Cyclization)

isomers were calculated by the ratio of the two methyl peaks at ~2.35δ and ~2.40δ, which

72

integrate together to three hydrogens, as shown in **Figure 28**. As the derivatives of these

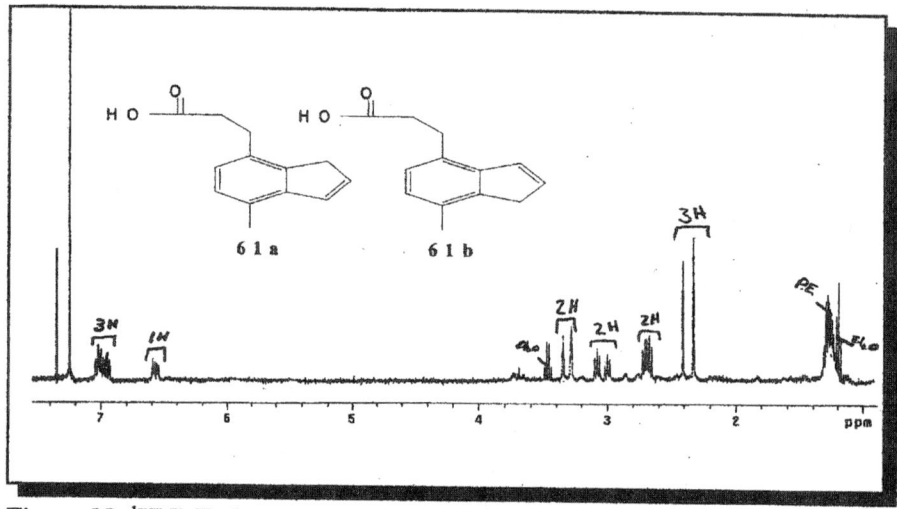

Figure 28 ¹HNMR Spectrum of 3-(inden-1yl)propanoic acids

isomers form the same indenylate when deprotonated, it was not necessary to separate them.

In fact, even isolating these acids proved somewhat difficult. When the actual acids were isolated, the yields were fairly low (around 20%). However, if the crude acids were reacted with a large excess of lithium aluminum hydride, much higher yields of the alcohol (calculated from the diketone or even the 2-ketoglutaric acid) were obtained. This fact suggested again that the problem is not with the reaction, but with the isolation in the work up.

The fact most indicative of the reaction's actual success is that when neither initial acid (4,7-dioxooctanoic acid or indenyl propanoic) was isolated, but each crude mixture reacted immediately in the next step (**Figure 29**), much higher yields resulted. In

73

the overall reaction sequence
from 2-ketoglutaric acid to
isomeric alcohols **62a** and **62b**
yields up to 92% on the one
gram scale and up to 79% on the
ten gram scale were obtained

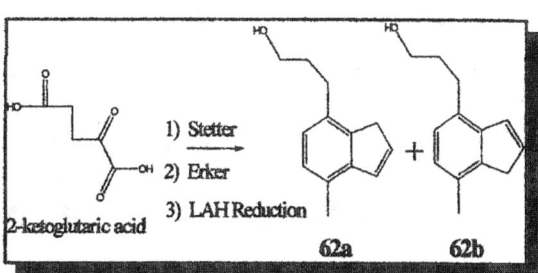

Figure 29 Reaction Sequence Followed Sans Isolations

(though yields of 60% were more common).

From the
isolated propanoic
acids, the lithium
aluminum hydride
reduction gave up to
94% yield of alcohol

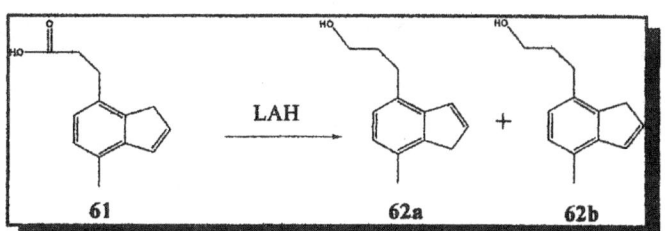

Figure 30 Alcohol **62** from isolated indanoic acid **61**

34 [129]. However, with the 20% yield after isolating the propanoic acid mixture and the

40 % yield after isolating the 4,7-dioxooctanoic acid (or 70 +% by the Novak method),

the overall yield of a good reaction sequence from 2-ketoglutaric acid to alcohol **62** (**a**

and **b** isomers) was 7 or 8 %, if each step was isolated. Since the using the crude

mixtures gave up to 92% yield of alcohol **62** (**a** and **b**) from the initial 2-ketoglutaric acid,

the 4,7-dioxooctanoic acid and 3-indenyl propanoic acids must be formed in near

quantitative yield. The far less than quantitative yields in the individually isolated steps,

therefore, must be due to loss in the workup and isolation.

Since the methodology of generating alcohol **62** in high yield bypassed the

74

isolation of the intermediate acids, we felt it necessary to complete a full characterization

of this compound to establish conclusively that we had indeed achieved the compound.

To this end, an attempt to separate the double-bond isomers was made. An exceptionally

long column with petroleum ether as eluent was able to give fractions with each isomer

significantly enhanced over the other. This process allowed all carbons and hydrogens to

be assigned based on a comprehensive NMR analysis that included ^{1}H, ^{13}C, HMBC,

HMQC, COSY, and NOESY spectra.

The first isomer to pass through the column was designated **a** while the trailing

isomer was designated **b**. The comprehensive NMR analysis allowed the determination

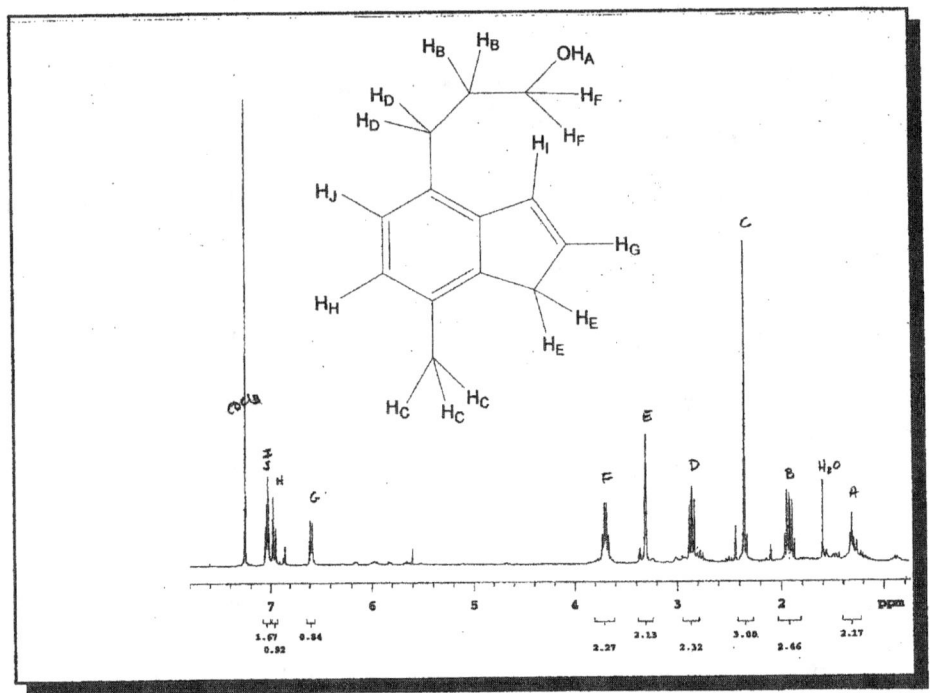

Figure 31 Alcohol **62a**, First Isomer through Column

75

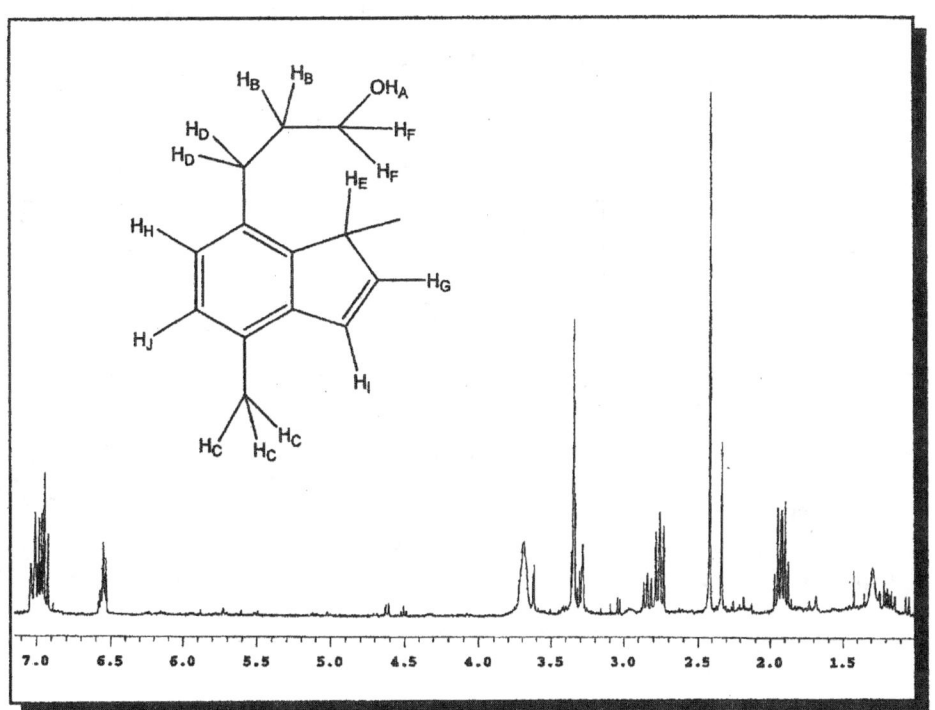

Figure 32   Alcohol **62b**, Second Isomer through Column

of the structure of each as shown in **Figures 31** and **32**.

The assignment of hydrogens to peaks was relatively straightforward. Alcohol

62a was slightly more pure than alcohol **62b**. Therefore, its structure was solved and the

structure of **62b** mostly inferred from the exhaustive assignments performed on **62a**. This

was accomplished by first assigning the hydrogens in its ^{1}H NMR spectrum labeled A-J

for each set of hydrogen signals from lowest to highest shift. The hydrogen sets are given

in **Table 1**. Based only on the proton spectrum, the isomer could not be absolutely

determined. However, the fact that the methyl group (H_C) was more shielded in **62b** than

in **62a** , and the reverse for the methylene nearest the ring (H_D) suggested that the double

76

bond was closer to the methyl in **62b** and the methylene in **62a**.

A COSY spectrum (**Figure 33**) was obtained to confirm the couplings assumed by the coupling constants and in hopes that the

Hydrogen Set	Chemical Shift	Integration	Multiplicity	Coupling Constants
A	1.292δ	1H	t	J_{AF} = 7.0 Hz
B	1.899δ	2H	tt	J_{BD} = 10.0 Hz J_{BF} = 8.5 Hz
C	2.334δ	3H	br s	
D	2.839δ	2H	br t	J_{DB} = 7.0 Hz
E	3.287δ	2H	dt (looks t)	J_{EI} = 3.0Hz J_{EG} = 3.0Hz
F	3.679δ	2H	dt(looks br q)	J_{FB} = 8.5 Hz J_{FA} = 7.0 Hz
G	6.568δ	1H	dt	J_{GI} = 7.5 Hz J_{GE} = 3.0 Hz
H	6.941δ	1H	br d	J_{HJ} = 10.5 Hz
I	7.015δ	1H	dt	J_{IG} = 7.5 Hz J_{IE} = 3.0 Hz
J	7.024δ	1H	br d	J_{JH} = 10.5 Hz

Table 1: Hydrogens for Alcohol **62a**

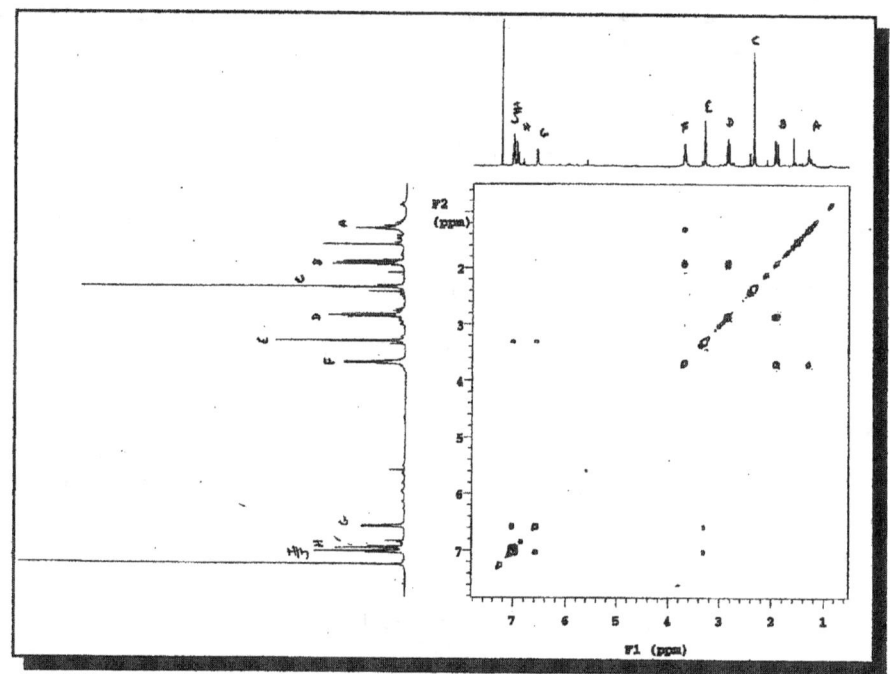

Figure 33 COSY Spectrum of Alcohol **62a**

77

coupling responsible for the broadening of the methyl peak could be determined, possibly confirming the position of the double bond. While the couplings were found to be correctly assigned (A to F, B to D and F, D to B, E to G and I or J, F to A and B, G to E and I or J, H to I or J, I/J to G and H), the COSY could not confirm the suggestion that the double bond was nearer the methylene than the methyl in alcohol **62a**.

To absolutely confirm this suggestion required an NOE measurement performed by a NOESY experiment. Irradiating the methyl group (H_C) in **62a** was expected to cause a nuclear Overhauser effect (through-space affect) on H_H and H_E while **62b** was expected to show its enhancement at H_H and H_J. As may be seen in **Figures 34** and **35**, this was the effect we found. Had the effects been reversed, then the opposite assignment would have been proven.

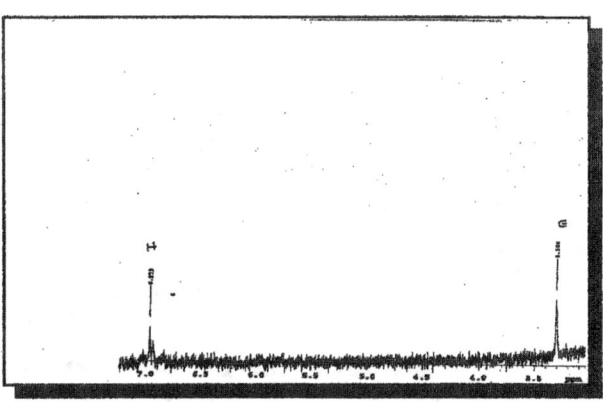

Figure 34 NOESY Spectrum of Alcohol **62a**'s Methyl Group

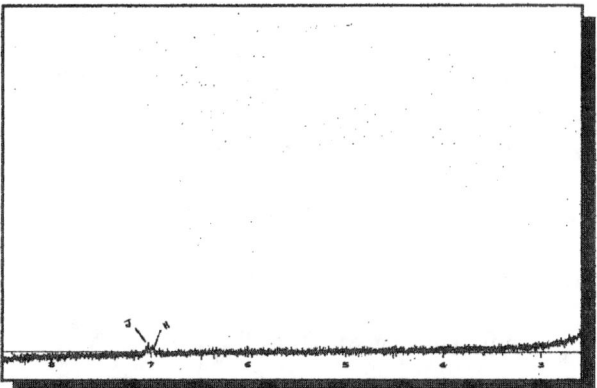

Figure 35 NOESY Spectrum of Alcohol **62b**'s Methyl Group

78

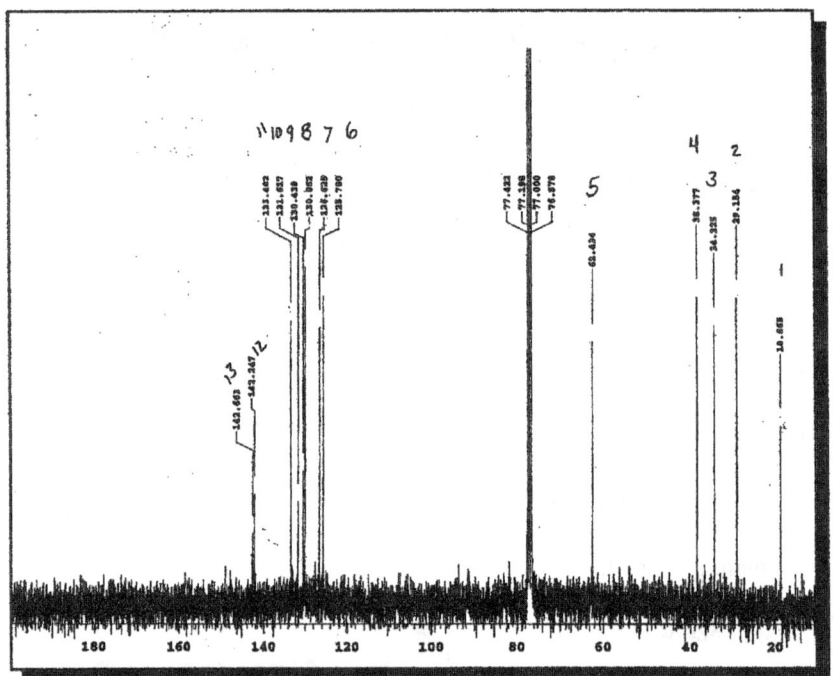

Figure 36 Alcohol **62a**'s ^{13}C Spectrum

A similar methodology was employed in order to concretely assign the carbons in alcohol **29a**. The basic ^{13}C NMR spectrum (**Figure 36**) was obtained and the peaks from each carbon sequentially assigned numbers from the lowest shift to the highest shift. This spectrum was then compared to the ^{13}C DEPT, HMQC, and HMBC spectra to ultimately get the assignments as shown in **Figure 37**. The DEPT spectrum

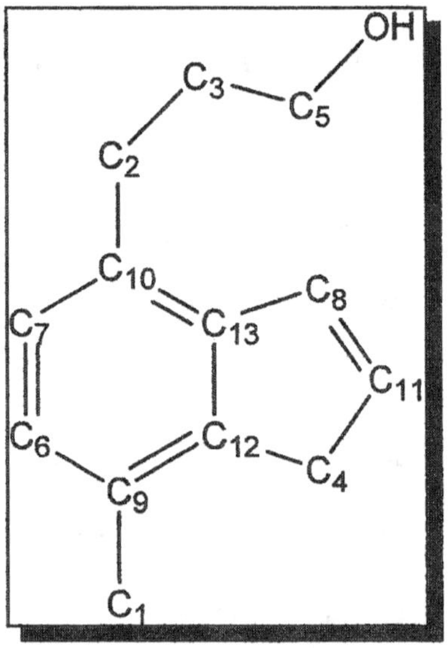

Figure 37 Carbon Assignments for alcohol **62a**

79

showed that the C1 was a methyl group, so it was assigned to the only methyl group in our compound. C2-5 were found to be methylenes, with C5 being in at a chemical shift indicative of an oxygen-bearing carbon, so it was assigned to our oxygen-bearing carbon. Carbons 6-8 and 11 are methines. The four signals from the original ^{13}C NMR spectrum that did not appear in the DEPT spectrum must be quaternary carbons (9, 10, 12, 13). The HMQC spectrum indicated that carbon 1 contained hydrogens C, 2 D, 3 B, 4 E, 5 F, 6 H, 7 I or J, 8 I or J, and 11 G. This information allowed the further assignments of carbons 2, 3, 4, 6, and 11. With the HMBC spectrum, the rest of the assignments were made. Table **2** contains a succinct summary of the information gleaned.

Carbon Number	Chemical Shift	DEPT	HMQC Coupled	HMBC Coupled
1	18.665δ	CH$_3$	C	H[#]
2	29.154δ	CH$_2$	D	B, F[#]
3	34.325δ	CH$_2$	B	D, F
4	38.377δ	CH$_2$	E	G[#]
5	62.434δ	CH$_2$	F	B, D
6	125.780δ	CH	H	C
7	126.625δ	CH	I or J	D
8	130.062δ	CH	I or J	C[?], E
9	130.439δ			C[?], G[#]
10	131.617δ			B[#], D, H[#]
11	133.402δ	CH	G	E
12	142.247δ			C[?], E, H[#]
13	142.663δ			C[?], D, I or J

#: weak
?: ancer data

Table 2: Alcohol **62a**'s Carbon Data

Carbon 12 and 13, being quaternary and having the largest chemical shift, were assumed to be the bridgehead carbons. Since carbon 12 couples to hydrogen E, it was assigned to the bridgehead position nearer the methyl group while carbon 13 with its coupling to the hydrogen D allows its assignment as the bridgehead nearer the methylene. Since carbon 10 couples

80

with hydrogen B, it must be the aromatic carbon bound to the methylene and the last quaternary carbon, carbon 9, must be the one bound to the methyl group. As this leaves only carbons 7 and 8 unassigned, carbon 7, with its coupling to hydrogen D and carbon 8, with its coupling to hydrogen E, must be the last aromatic carbon and the last alkenyl carbon, respectively. Thus, all hydrogens and carbons have been assigned.

The mass spectrum exhibited the molecular ion, fortuitously, as the base peak at 188.0. Typical of alcohols, loss of water is a significant daughter ion.

This completely characterized key intermediate compound is only mesylation and indenylate displacement thereof away from ligand 64 (**Figure 38**). Having the structure totally elucidated enables easy comparison with the ligand soon to be formed.

To this point, the synthesis typically used the simplified one pot, three reaction

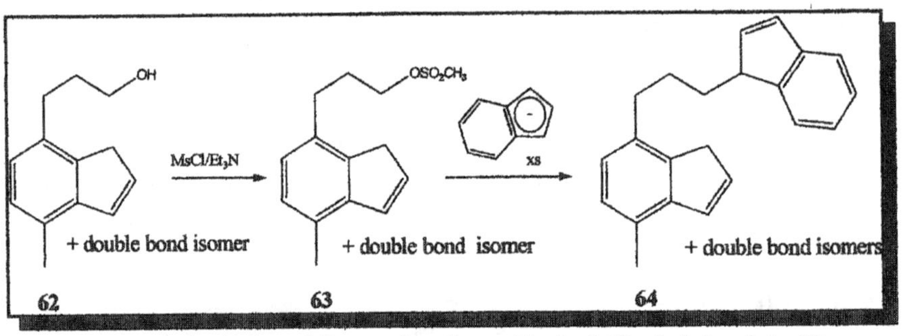

Figure 38 Typical Ligand Formation from Alcohol **62** via Mesylate **63**

method without the intermediates being isolated. Similarly, the mesylate of alcohol **62** (Compound **63**) was generally not isolated but formed *in situ* and then displaced by the indenylate to form ligand **64** (**Figure 38**). However, for completeness sake **63** was isolated and characterized.

81

The mesylate formed from alcohol **62** by methanesulfonyl chloride (mesyl

chloride or MsCl) and triethyl amine was actually quite stable and easily isolated. The ^{1}H

NMR spectrum, not surprisingly, looks much like the spectrum of the alcohol. The two

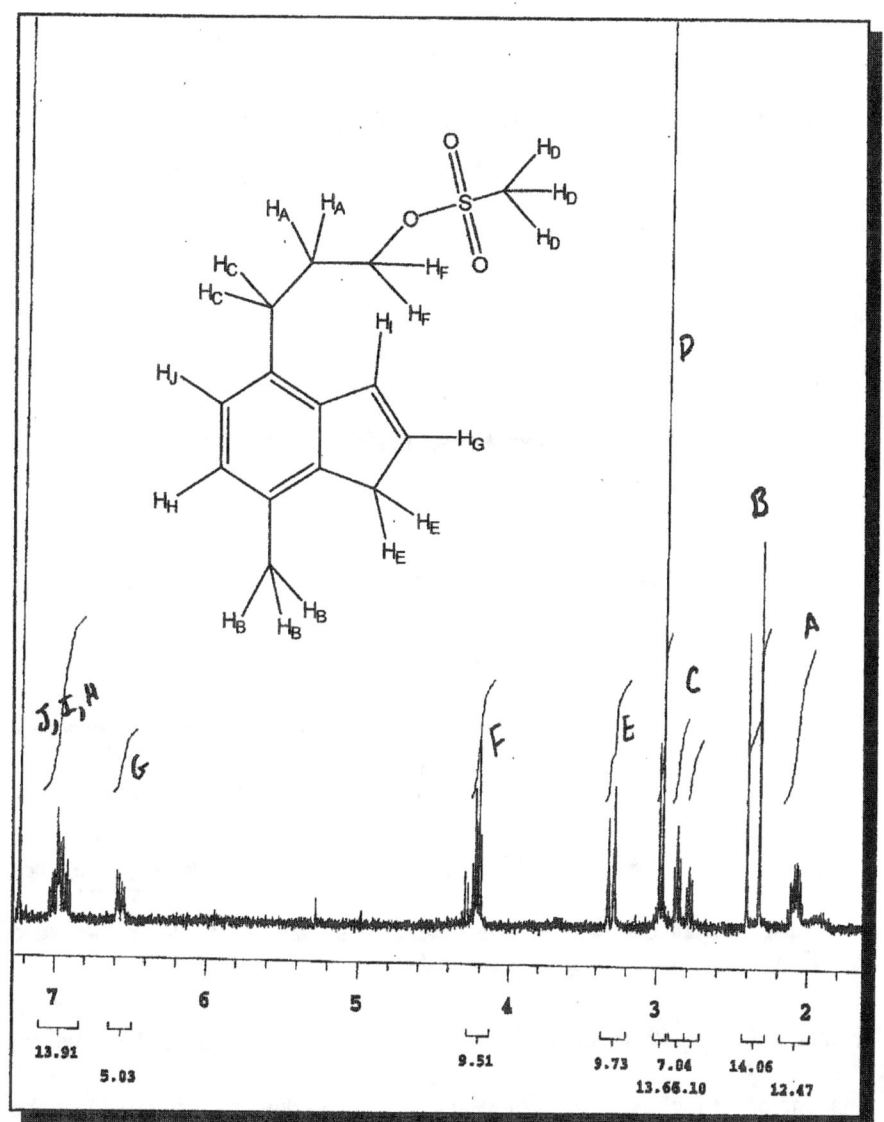

Figure 39 Mesylate **63**'s ^{1}H NMR Spectrum

82

major differences, as may be seen in Figure **50**, are the shifting of the F protons to up above 4.0δ and the replacement of the alcoholic hydrogen with a methyl singlet at ~3.0δ corresponding to the methyl on the sulfur. Otherwise, the spectrum is essentially unchanged.

The ^{13}C NMR spectrum is also highly analogous, except for the addition of a peak at ~38δ, which the DEPT spectrum (**Figure 40**) proves to be a methyl. Again, as this

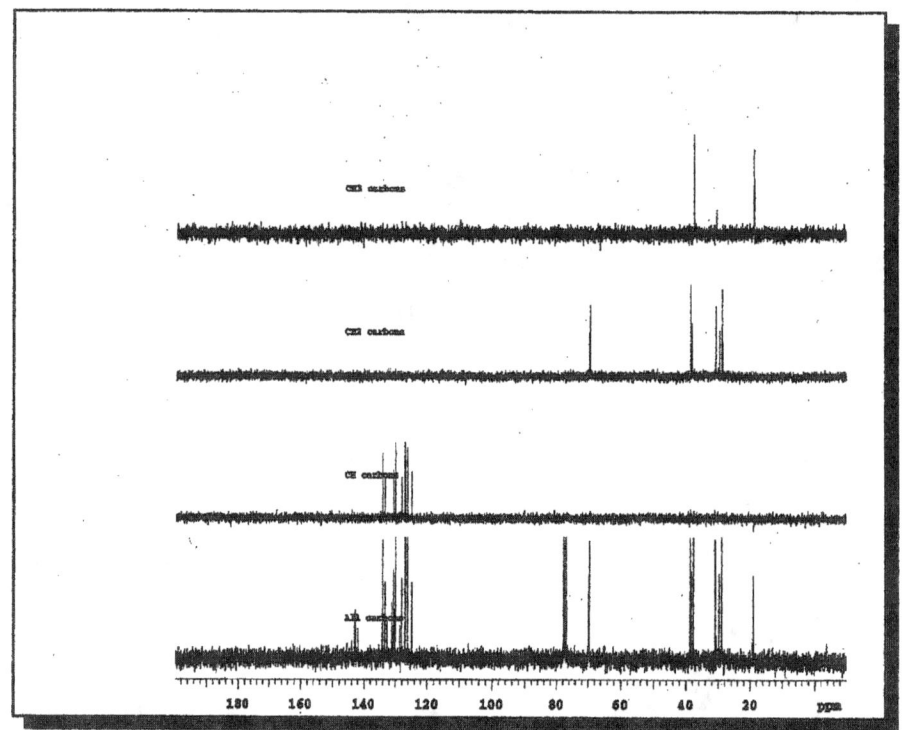

Figure 40 DEPT Spectrum of Mesylate **63**

compound was quite similar to the exhaustively determined alcohol, no real attempt was made to prove the assignments of each individual carbon and proton. Once the basic NMR spectra appeared consistent and the parent ion was confirmed in the mass spectrum,

83

a high resolution mass spectrum (HRMS) using Fast Atom Bombardment (FAB) and hydrogen as the carrier gas, was obtained. The MH$^+$ mass of 267.1055 (MH$^+$ calculated to be 267.1057) was found for a formula of $C_{14}H_{19}O_3$. With this confirmation, the mesylate was thereafter no longer isolated but immediately taken on to ligand **64**.

It might be noted that while the mesylate was displaced by the indenyl anion (indenylate) in this synthesis, the cyclopentadienyl or fluorenyl anions might also have been used. Again, our driving goal was the unique conformation of the 7-1'-bis(indenyl)

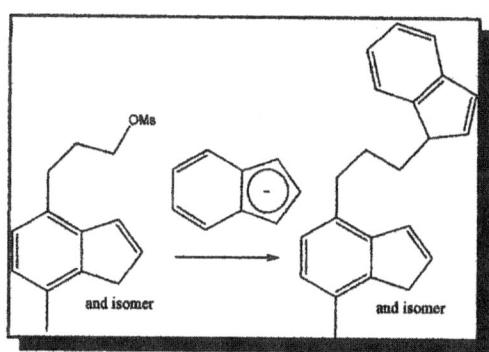

Figure 41 Ligand **64** Formation

propylene-bridged zirconocene, so the indenyl anion was chosen, though fluorenyl or cyclopentadienyl anions might also give interesting metallocenes.

The resulting ligand (1-(4-methylinden-7-yl)-3-(inden -1-yl)propane (Compound **65**) [and its double bond isomer (1–7-methylinden-4-yl)-

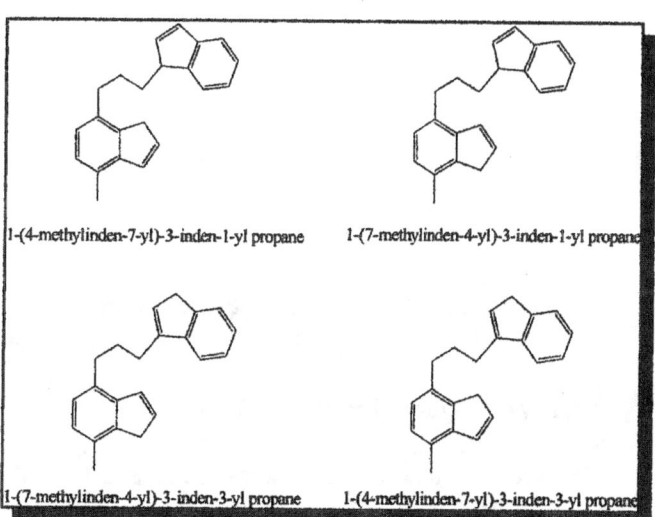

1-(4-methylinden-7-yl)-3-inden-1-yl propane   1-(7-methylinden-4-yl)-3-inden-1-yl propane

1-(7-methylinden-4-yl)-3-inden-3-yl propane   1-(4-methylinden-7-yl)-3-inden-3-yl propane

Figure 42 Ligand **64** Possible Double Bond Isomers

84

3-(inden-1-yl)propane] formed in good yield (generally about 70% crude, 45% purified, though one run actually went quantitatively). While the other inden-3-yl isomers shown in **Figure 42** are not formed directly in the reaction, the 1,3-hydrogen shift is well known in indenes and the more substituted, hence more stable, isomers could form. However, the ^{1}H NMR spectrum (**Figure 43**) shows only two significant isomers (based on the

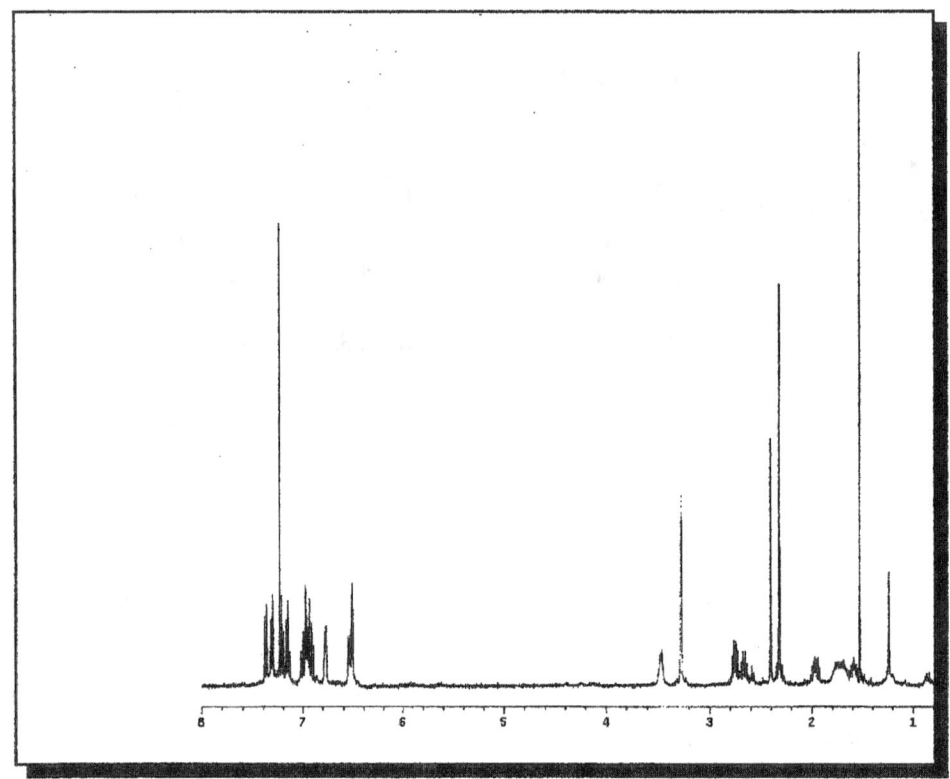

Figure 43 7,1'-Propylene-bridged bis(Indenyl) Ligand **64**

easily distinguished methyl peak). Either, neither of the double bonds shifted or both shifted completely. The answer lies in the fact that there is a hydrogen peak at 3.49δ that integrates to one hydrogen. If the inden-3-yl species were formed, this peak would

85

integrate to two hydrogens. Also, there would be one less alkenyl hydrogen in the inden-3-yl spectrum. The two originally formed isomers are the major isomers. A sample left for six months on the bench top showed significant broadening of the peak at 3.49δ and an decrease in its integration, as well as significant broadening of the two methyl peaks, suggesting that the double-bond isomerization is occurring, but very slowly. However, since the metalation involves deprotonation of both indenyl rings, all the possible isomers would give the same dilithium complex (**Figure 44**) and absolute determination of which isomers are present at metalation was actually unnecessary.

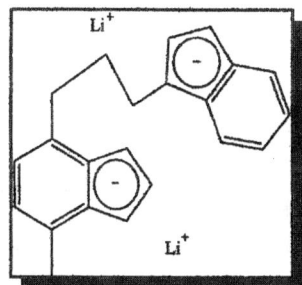

Figure 44 Dilithium Complex of Ligand **64**

The yield of the ligand, whether formed by *in situ* formation and displacement of the mesylate (one pot) or isolation of the same (two pot), was relatively unchanged. The isolated yield for just the mesylation was generally in excess of 92% and the S_N2 displacement by the indenylate up to 68%. However, using the isolated mesylate resulted in significantly less indenylate being required and, thus, was less difficult to separate from the excess indene after completion of the reaction and the final purified yield around 40-50% from the alcohol, quite comparable with the in situ result.

The basic ^{1}H NMR spectrum (**Figure 43**) of ligand **64** showed four new aromatic hydrogen signals and loss of the signals from H_F (**Figures 31** or **39**, depending on

86

whether the one or two pot method, respectively, was used) as would be expected.

Expanding the spectrum significantly and comparing to the spectra obtained for alcohol

62, allowed the hydrogens to be assigned as shown in Figure **45** and **46**.

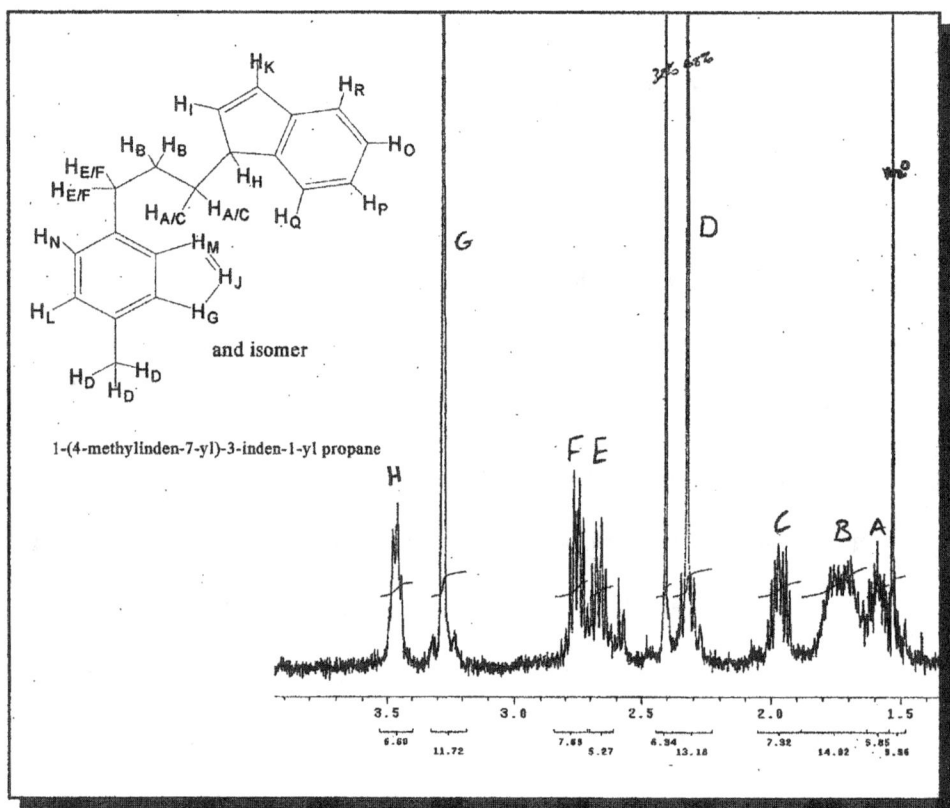

Figure 45 Ligand **64** Expanded Alkyl Region

The presence of both isomers [(1-(4-methylinden-7-yl)-3-(inden-1-yl)propane and

its double bond isomer 1–(7-methylinden-4-yl)-3-(inden-1-yl)propane] made assignments

a little difficult. However, by measuring the ratio of the two isomer's distinct methyl

peaks (32:68 in this run) and comparing ratios of the overlapping regions of each

87

corresponding isomers hydrogen systems, a good degree of certainty can be obtained.

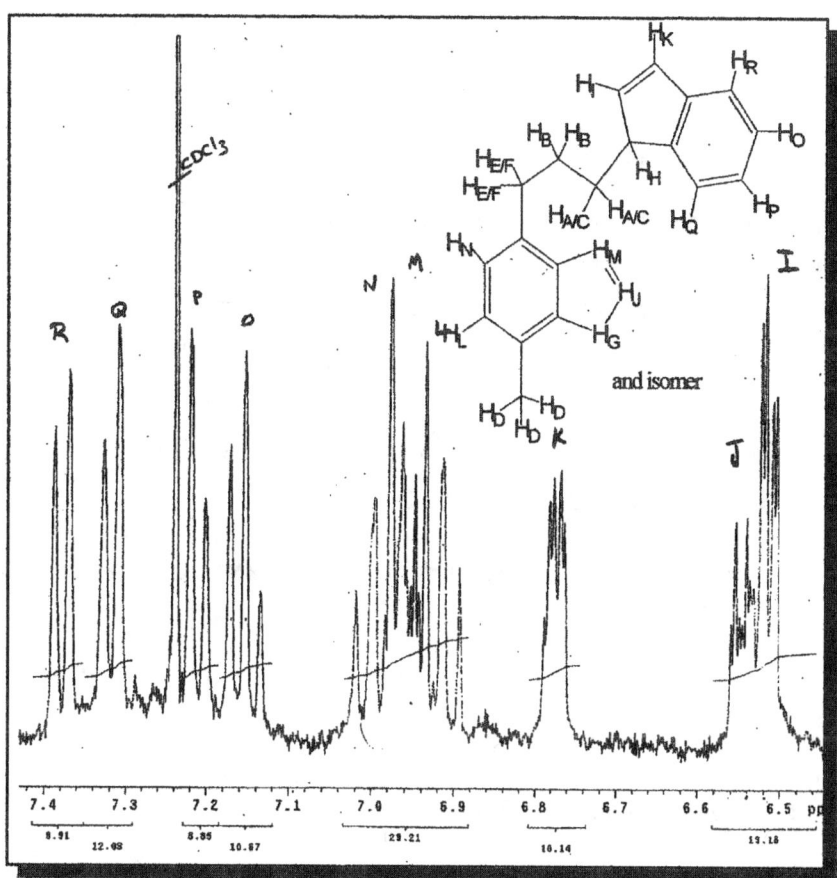

Figure 46 Ligand **64** Expanded Aromatic Region

A simple mass spectrum was also obtained which showed the molecular ion peak as a significant peak while the loss of the methylated indene system accounted for the base peak. The high resolution mass spectrum (HRMS) using Fast Atom Bombardment (FAB) and hydrogen as the carrier gas, gave an MH$^+$ mass of 287.1800 for a formula of $C_{22}H_{23}$.

88

The presence of the two inseparable isomers of the ligand made complete assignment of carbons impossible. Although all peaks are seen in the ^{13}C NMR spectrum, absolute assignment of which peaks correspond to which isomer was not possible.

The 7,1'- propylene-bridged-bis-indenyl ligand was metalated through typical procedures (**Figure 47**) [102, 103] to achieve the ansa-propylene-bridged 1-(7-methylinden-4-yl)-3-inden-1-yl zirconium(IV) dichloride **65** in good yield (generally from

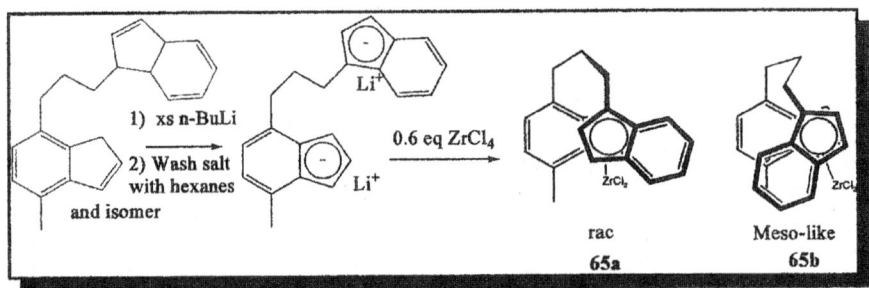

Figure 47 Typical Metalation of bis(Indenyl)Ligands

~60% to ~80%) in approximately a 60:40 ratio of stereoisomers. The ratio of the stereoisomers was determined by the integration of the very obvious 3H singlets corresponding to the methyl group of each isomer and the remaining lithium salt of the ligand (**Figure 48**).

While the hope had been for a single conformation of a single diastereomer, this was not the case.

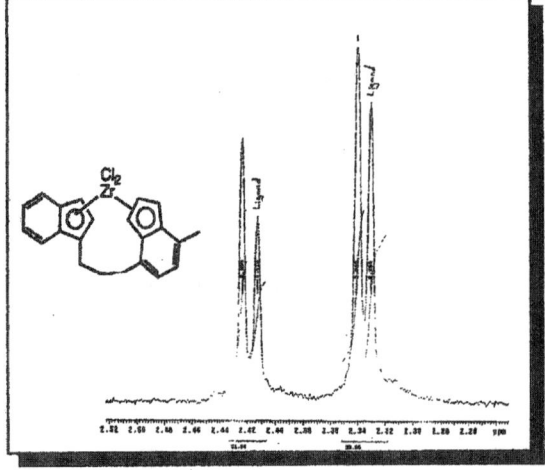

Figure 48 Methyls of **65** and dilithium salt of **64**

89

Other attempts at metalation by varying conditions such as metalation solvents and temperatures were unable to change the ratio significantly. Recrystallization procedures were able to remove the lithium salts (**Figure 49**). However, recrystallization attempts

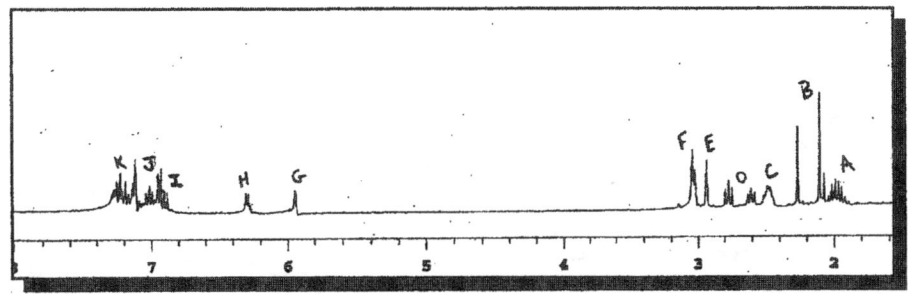

Figure 49 ^{1}H NMR Spectrum of purified Zirconocenes **65a** and **b**

were not sufficient for the isolation of the individual isomers, which remained aloof.

Repeated attempts at crystallization finally succeeded in growing long, fine crystals – too fine for x-ray crystallography. It might be noted that in all the attempts at metalation the same isomer (with the lower shifted methyl) was the major isomer. In the meso-like isomers, the inden-4-yl's methyl group would be significantly closer to the inden-1-yl's aromatic ring than in the rac-isomers and can account for the observed shift.

However, this is a rather tenuous argument and without an x-ray crystallographic structure we did not feel comfortable assigning the peaks to a specific isomer structure on this evidence.

In multiple metalation attempts we noticed that the ratio of the stereoisomers corresponded to the ratio of the isomers of the

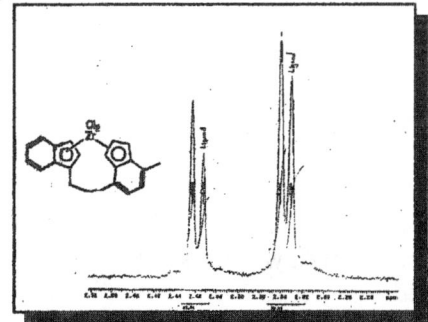

Figure 50 Ratios of **65** and dilithium salt of **64**

90

dilithium salt of the ligand, exemplified here in **Figure 50**. This rather suggests that the stereochemistry is set in the dilithium salt and that the lithium binds tightly to the face it originally binds and is only removed by the zirconium's approach with either complete retention or complete inversion. The two cyclopentadiene units each have two faces and so four possible dilithium complexes can result. However, as may be seen in **Figure 51**, if the planar chirality is reversed by binding the opposite face for both cyclopentadiene moieties, a pair enantiomers results. Since

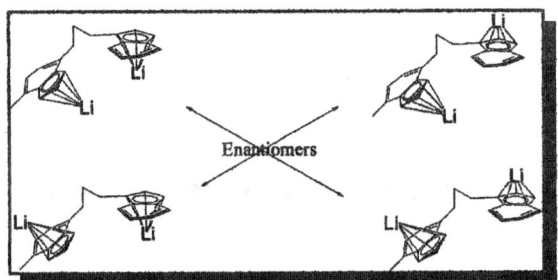

Figure 51 Possible Dilithium Complexes – 2 Pairs of Enantiomers

enantiomers have the same physical properties, only two sets of peaks should show in the ^{1}H NMR spectrum. In fact, the pair of peaks are seen as expected (**Figure 50**).

However, one of these peaks is at 2.33 δ and the other at 2.41 δ, a large shift for a change in a stereocenter nine carbon units away. To account for this shift, we suggest that the lithium must be forming some sort of aggregate, perhaps with a molecule(s) of the solvent or through chlorine(s). This bridging effect holds the dilithium complex in a rac-like and meso-like position (**Figure 52**). The methyl group would then be near the inden-1-yl ring in the meso-like and farther away in

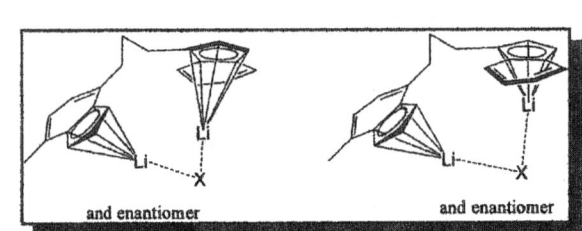

Figure 52 Proposed Aggregation of Lithium Salts

91

the rac-like, accounting for the large difference in the shift of the methyl groups.

If this is indeed the case, then perhaps forming the potassium indenylate of the ligand would allow greater selectivity in the metalation. After metalation, even refluxing the metallocene in toluene does not change the ratios. The fact that the ratios do not change even in refluxing toluene suggests that there is no interconversion between the two isomers formed. This strongly suggests that the two zirconocene isomers represented by these two peaks are the meso-like and d,l enantiomeric pairs.

In **Figure 53** the theoretical metalation of the four possible dilithium salts is shown with complete inversion. If the lithium directed the incoming zirconium, then

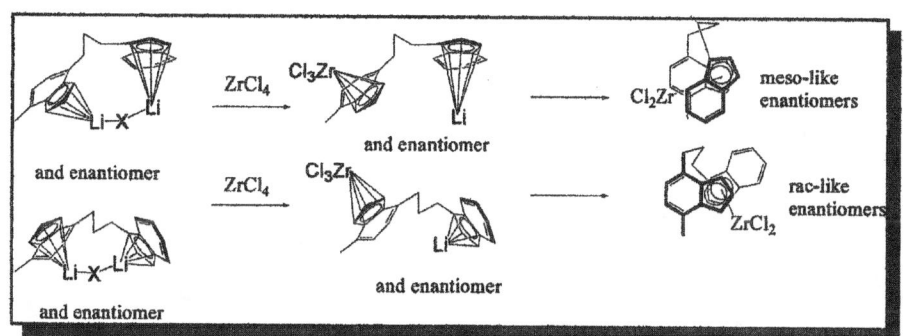

Figure 53 Theoretical Metalation of Dilithium Salts of Ligand **64**

complete retention would occur and the enantiomer of each final metallocene would be formed. Either possibility, complete retention or complete inversion, would account for the observation of the ratios of the lithium salts correlation to the ratios of the zirconocene isomers formed.

As previously mentioned, the suggestion that the lithium salts are forming non-interconvertible isomers is born out by the ^{1}H NMR spectrum (**Figure 50**). Absolute

92

confirmation of this suggestion came with the formation of the dilithium crystals of the ligand with a large excess of n-butyllithium in hexane (**Figure 54**). This allowed the confirmation of the peaks believed to belong to the methyl's of the two dilithium isomers. Therefore, it becomes likely that the two additional peaks after metalation are the meso and *d,l* (rac) stereoisomers.

The ^{1}H NMR spectrum helps to confirm the presence of the *d,l* (*rac*) and *meso-*

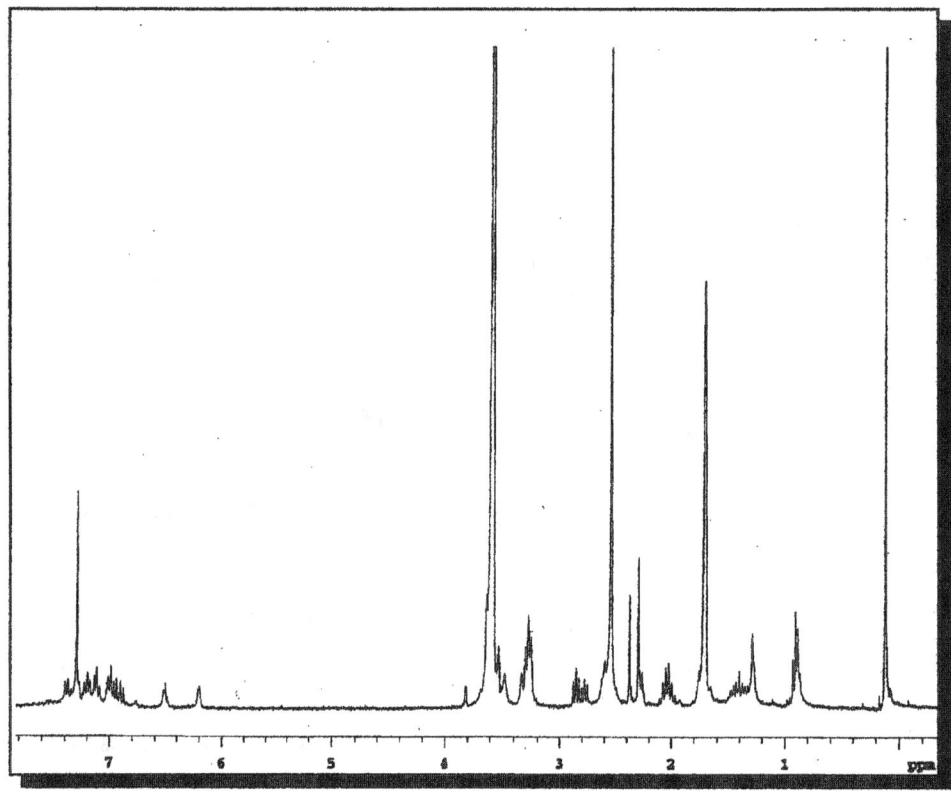

Figure 54 Dilithium Crystals of **64** and n-butyllithium.

like stereoisomers. While the isomers were unable to be separated enough to completely

93

assign the individual protons, reasonable assumptions may be based on the ^{1}H NMR and

COSY spectra with reference to the expected geometries of the two isomers. In the major

isomer, the B hydrogens couple to the I, J, and F hydrogens while in the minor isomer the

B hydrogens couple to E, F, and I. As may be seen by the expected geometries in **Figure

55**, two of the cyclopentadiene hydrogens are eclipsed by the benzene ring in the Rac-

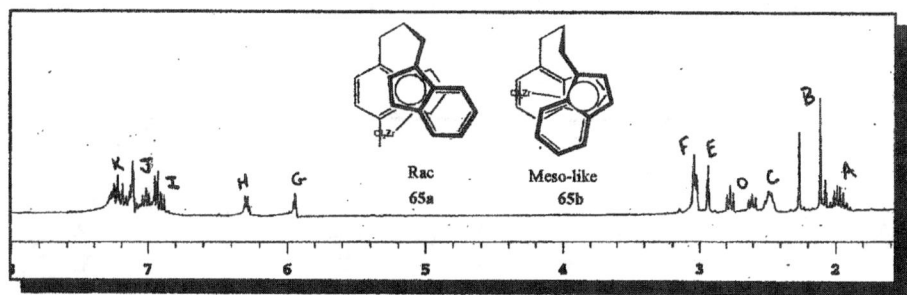

Figure 55 ^{1}H NMR of purified Zirconocenes **65a** and **b**

isomers whereas only one is eclipsed in the *meso*-isomer. Since two of the hydrogens (E

and F) on the inden-4-yl's cyclopentadiene unit in the minor structure are shielded by the

other benzene ring, as evidenced by their occurrence at shifts near 3.0 δ and only one

hydrogen in the major structure is so shifted, it is reasonable to assume that the major

isomer is the *meso*-like isomer and the *rac*-isomer is the minor isomer.

We desired a single conformer of a single diastereomer and we obtained two

diastereomers. Obviously, the three-carbon chain allows too much mobility for the

isolation of a single conformational isomer of a single diastereomer. Also, it might be

worthy of note that the PC model calculations for the two zirconocenes that predicted the

meso-like diastereomers to have higher heats of formation was obviously in error as we

discovered the meso-like diastereomers to be the major component of the mixture.

94

4.4   **Ethylene-bridged-1-(7-Methylinden-4-yl), 2-(inden-2-yl)zirconium(IV)**

dichloride

The synthesis achieved in the previous section had, at its heart, a 3-(7methylinden-4-yl)-propanoic acid (and its double bond isomer). As mentioned there, the esterification of this acid (**61**) to ester **66**, followed by the known diGrignard addition to esters [131]

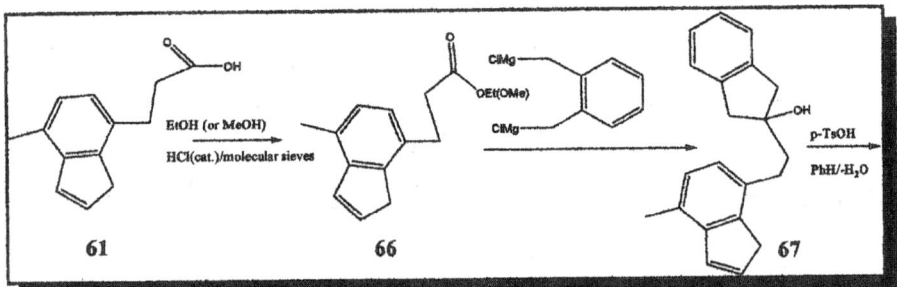

61   **66**   **67**

Figure 56 Planned Route to Ligand **68**

would form the 2-indanol species **67** which is only loss of water away from bis-indenyl ligand **68**. While the faces of this ligand are enantiotopic and unlikely to form conformational isomers on metalation, the enantiomeric zirconocene so formed should itself be of intermediate conformation (near II). It was therefore synthesized and metalated.

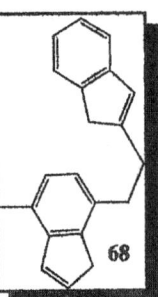

Figure 57
Ligand **68**

A crude sample of the acid was analyzed by ^{1}H NMR spectroscopy. Molecular weights and peak integrations were used to calculate the actual mass of acid. The sample was then refluxed for two weeks over copious molecular sieves

95

in absolute ethanol (purified by passage through basic alumina) with a few drops of

concentrated hydrochloric acid added as catalyst. This procedure produced ester **66** in

high yield (92%). Repeat runs without the molecular sieves in either ethanol or methanol

reached yields of only 60%, but were also stopped after only refluxing for a few days.

The ^{1}H NMR spectrum, shown in **Figure 58**, was very clean. Following the

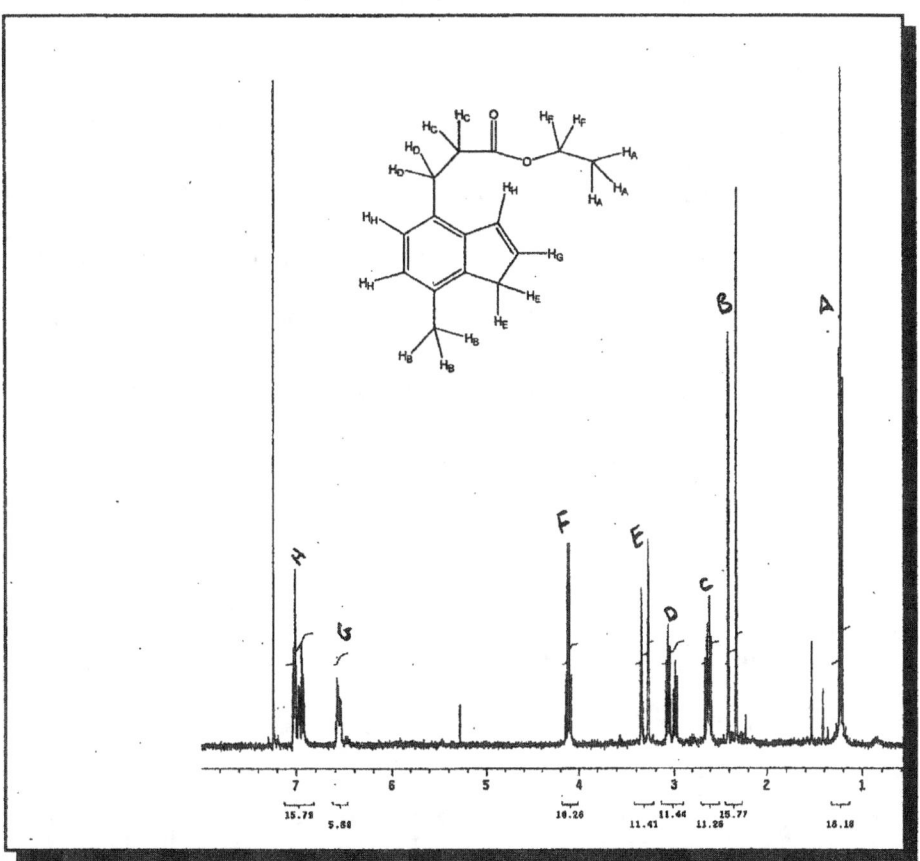

Figure 58 ^{1}H NMR Spectrum of Ester **66**

procedure established in the assignment of peaks in alcohol **62**, the spectrum was

assigned letters to each hydrogen system, A-H. Hydrogen system A at 1.25 δ, a 3H triplet

96

with 8 hertz coupling, was assigned to the methyl group of the ester while hydrogen

peaks F, a 2H quartet with 8 Hertz coupling corresponded to the methylene of the ester.

The peaks assigned the letter B, corresponding to two 3H singlets were assigned to the

remaining methyl group, the one on the ring. The hydrogens assigned to E are at the

typical shift for the 1 position of indene and were therefore assigned to the 1-position in

our indenyl compound. Hydrogens C and D must belong to the hydrogens of the side

chain. Based on assignments in the indanoic acid **61** and the similar alcohol **62** in the last

section where the corresponding hydrogen nearest the aromatic ring had a shift near 3, the

D was assigned to the hydrogens nearer the ring while C was assigned to the hydrogens

nearer the ester. Finally, the aromatic and alkenyl hydrogens were assigned based on the

exhaustively determined alcohol **62** wherein the hydrogen at the lower shift (~6.5 δ)

corresponded to the 2-position of the indene and the three hydrogens near 7.0 δ were

assigned to the ring hydrogens and the hydrogen at the 3-position.

The diGrignard reagent necessary to form the 2-indanol **67** from ester **66** was

formed precisely according to the method in the literature [131]. However, we found that the formation only worked successfully when the starting dihalide was the dichloride and was fractionally distilled immediately prior to use.

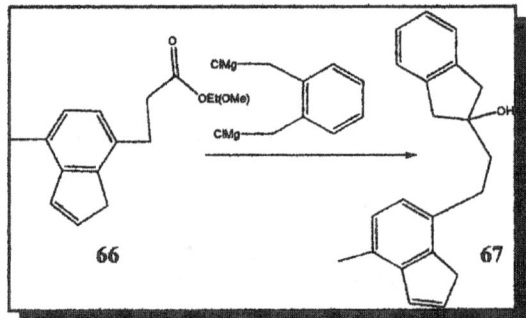

Figure 59 DiGrignard Reaction

Because of the high freezing point of the dichloride, a heated condenser tube had to be

97

used to keep the purified material from freezing in the condenser. However, once the dichloride had been freshly distilled, the diGrignard was relatively easily formed and ester **66** was added. When no trace of ester **66** remained on the TLC plates or ¹H NMR spectra of aliquots of the reaction mixture, the compound was worked up by typical procedures to give the resulting alcohol in essentially quantitative yield.

The loss of the ester peaks at 1.25 δ and 4.10 δ from the ¹H NMR spectrum (**Figure 60**) was quite obvious, indicating the consumption of the ester. Again, the

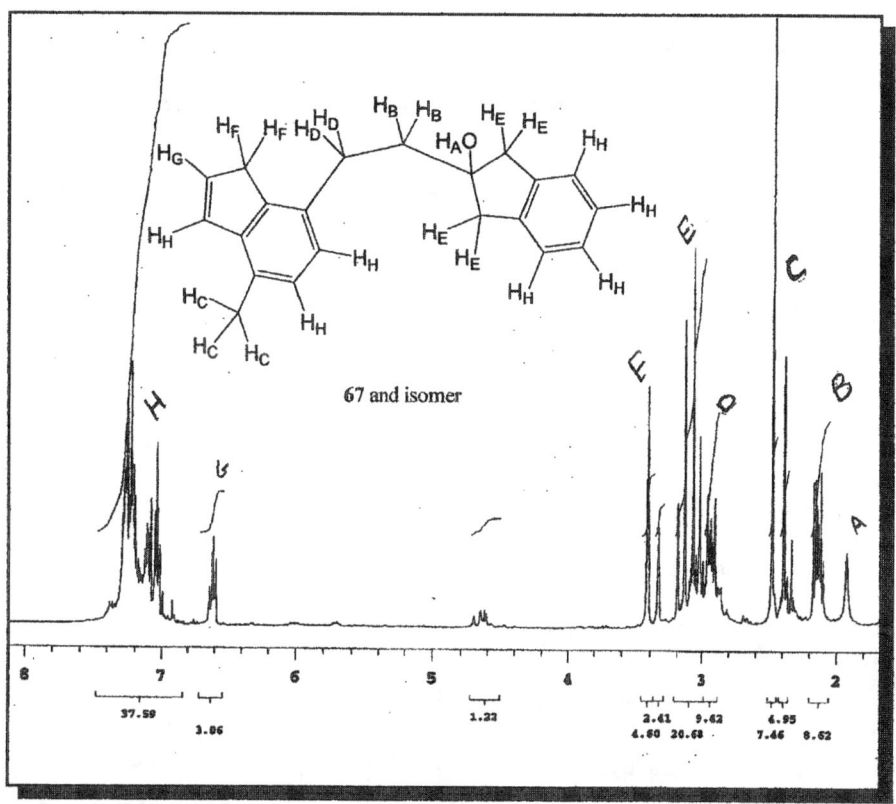

Figure 60 ¹H NMR Spectrum of Alcohol **67**

spectrum's hydrogen systems were alphabetically labeled from lowest to highest shift.

98

The lowest shift, hydrogen A, a broad singlet integrating to 1H, was assigned to the hydrogen of the alcohol. The next most obvious assignment was hydrogens C, two singlets that together integrate to 3H and correspond to the only methyl group in the isomers. Hydrogen D is a 2H multiplet just before 3δ and is assigned to the benzylic hydrogens due to the similarity to previously determined compounds. Similarly, the F hydrogens are assigned to the 1-position of the indenyl moiety by their similarity to the shift of indene's 1-position hydrogens. Hydrogens B, a 2H multiplet, were assigned to the bridgehead nearest the indanol as it is the most alkyl-like methylene in the molecule and should have the lowest shift. The single hydrogen G is assigned to the 2-position of the indene based on previously examined compounds. The H hydrogens overlapping the deutero-chloroform peak correspond to the seven aromatic hydrogens. That the integration truly is seven hydrogens was confirmed by repeating the spectrum with deutero-methylene chloride as solvent. The assumption of seven hydrogens was confirmed. Finally, the diastereotopic methylene hydrogens of the indanol were assigned to the hydrogens E,

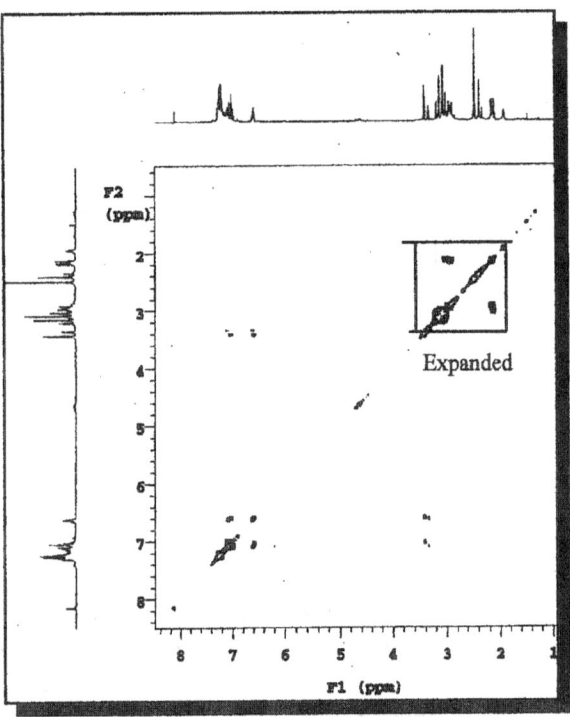

Figure 61 COSY Spectrum of Alcohol **67**

99

which integrate together to 4H. A COSY spectrum (**Figure 61**) showed the expected couplings between H_B and H_D and between H_F and H_G. An expansion of the 2-3δ section of the spectrum (**Figure 62**) also shows the coupling of the diastereotopic hydrogens in H_E with their geminal counterparts. With this confirmation, the proton assignments were assumed correct.

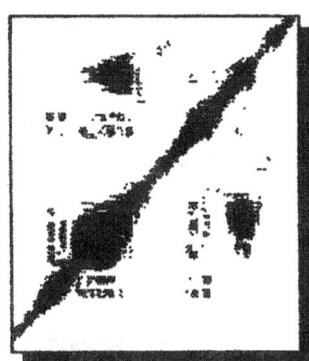

Figure 62 COSY Spectrum ~2-3δ

Since the isomers of alcohol **62** had been inseparable and the isomers of alcohol **67** should be even more difficult to separate, we did not attempt separation. This made exact assignments of the peaks in the ^{13}C NMR spectrum (**Figure 63**) difficult. However,

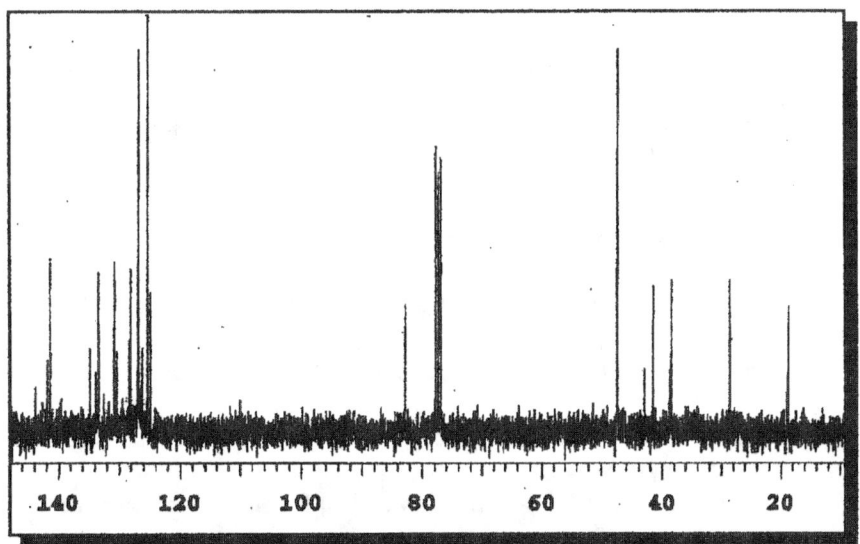

Figure 63 ^{13}C Spectrum of Alcohol **67** Isomers.

the spectrum was consistent in that approximately the right number of alkyl-like and aromatic-like signals were obtained. Also, a prominent peak at 80 δ, consistent with an

100

alcohol-substituted carbon, was observed. Along with the DEPT spectrum, **Figure 64,**

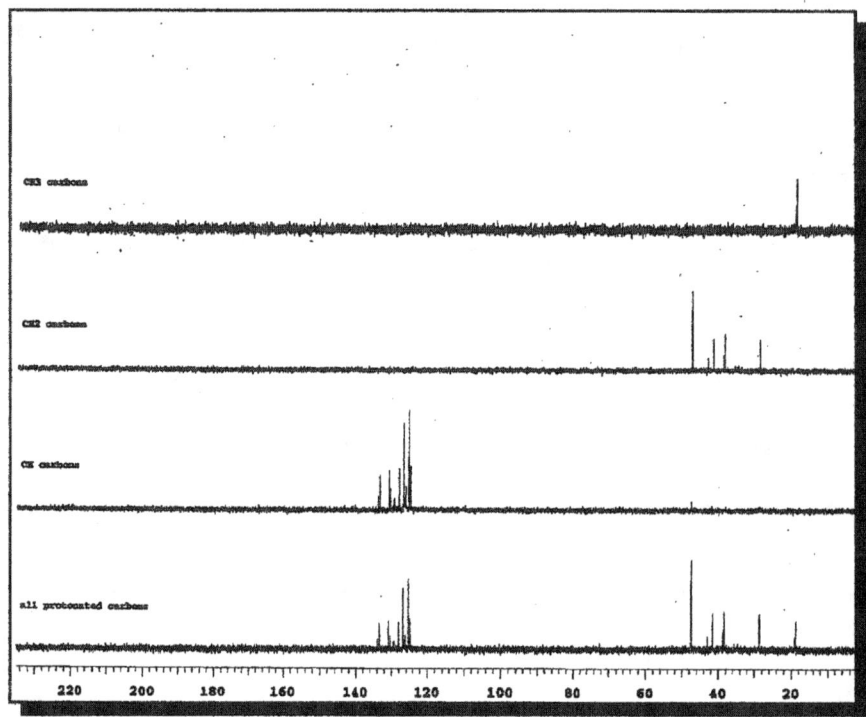

Figure 64 DEPT Spectrum of Alcohol **67**

and the absence of the quaternary alcohol's peak therein, we felt confident that the

compound was correct. Similarly, the DEPT spectrum shows a single methyl group and

five methylene in the alkyl region, further justifying our faith in the compound. A mass

spectrum obtained by electron bombardment (70 eV) gave a base peak of M - 18. This

corresponds to the loss of water, fairly typical for alcohols. Using the + Na technique

with high resolution mass spectroscopy, a spectrum was obtained with a mass of

313.1577 for the mass plus sodium compared to the calculated mass of 313.1568,

confirming the compound.

101

Alcohol **67** was then treated with p-toluenesulfonyl chloride and refluxed in the Dean-Stark apparatus with benzene for the azeotropic removal of water. The yield for this reaction was a poor 20% on the initial run. To make matters even worse, the other 80% of starting material was not recovered.

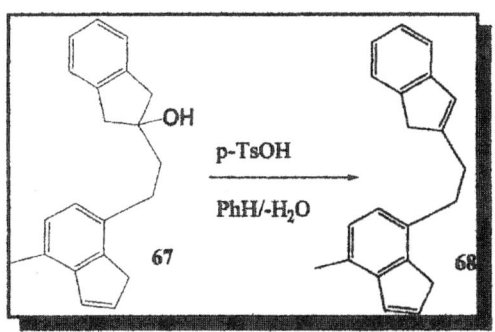

Figure 65 Dehydration of Alcohol **67**

However, the 20% yield of product had a mass of 192 mg and looked very pure by ^{1}H NMR (**Figure 66**, next page) after a single silica plug. Assignments for the protons were made by the now familiar alphabetical labeling from lowest shift to highest shift. Hydrogen A, integrating to 3H, obviously corresponds to the methyl group. Hydrogens B and C must be the bridging methylenes. Hydrogen C was assigned the position nearer the inden-7-yl ring based on previous spectra, so B must be the hydrogens on the carbon nearest the inden-2-yl ring. The hydrogen set D integrates to 4H and corresponds to the two 1-position indenyl hydrogens. Hydrogen set E was a bit puzzling as it integrated to 2H and is at a shift previously associated with hydrogens at the 2-position of an indenyl unit. However, if the presence of the bridge at the 2-position is assumed to shift the hydrogen from the nearby 3-position, then it can account for one of the 2H there while the actual 2-position hydrogen of the other indene can account for the second. The three hydrogen of set F are assumed to be the aromatic carbons of the inden-7-yl ring and the 3-position based on previous, similar spectra. Hydrogens G, H, and I are assigned by a

102

combination of the splitting patterns and the shifts expected based on the Silverstein predictions (176).

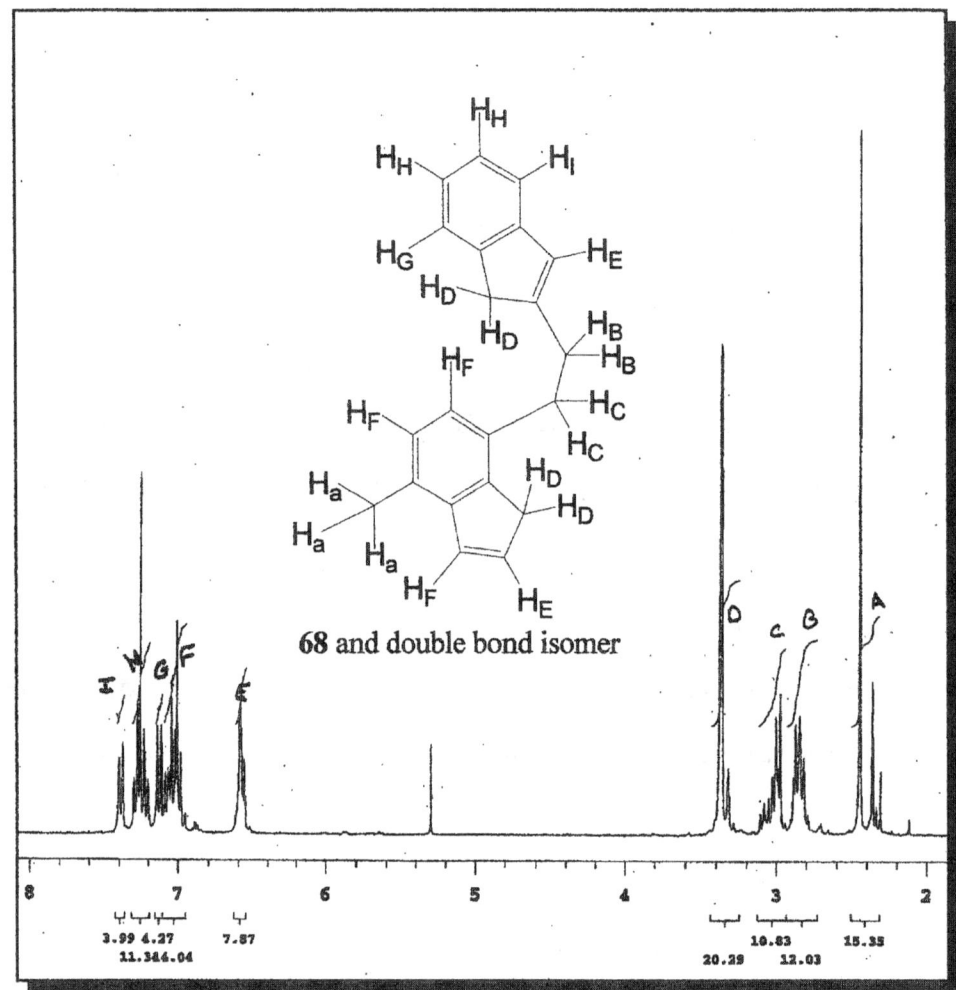

68 and double bond isomer

Figure 66 ^{1}H NMR Spectrum of Ligand **68**

After a preparatory thin layer chromatography (Prep TLC) separation of the isomers, the ^{13}C NMR spectrum was obtained (**Figure 67**) for one of the isomers.

103

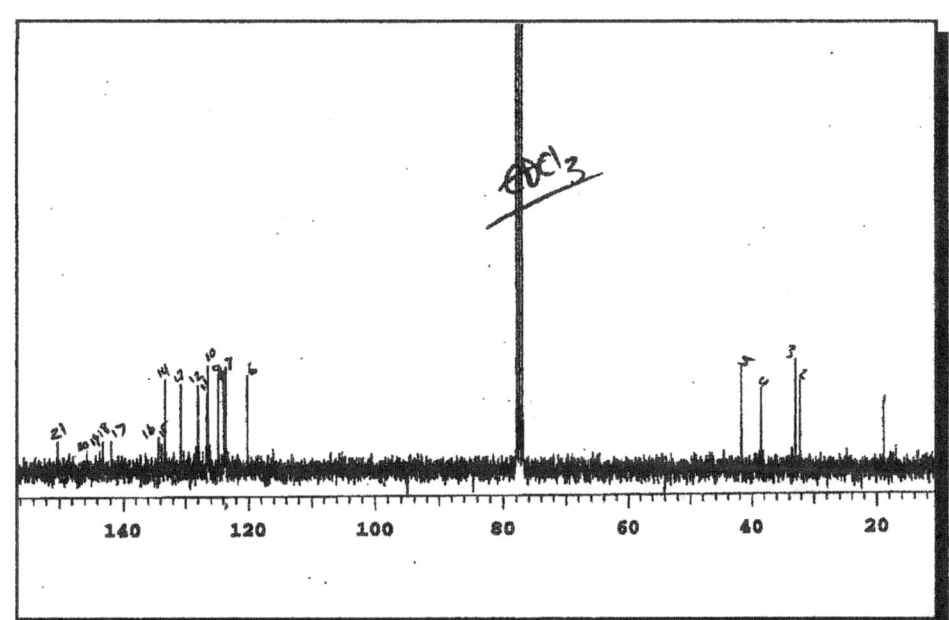

Figure 67 ¹³C NMR Spectrum of Ligand **68**

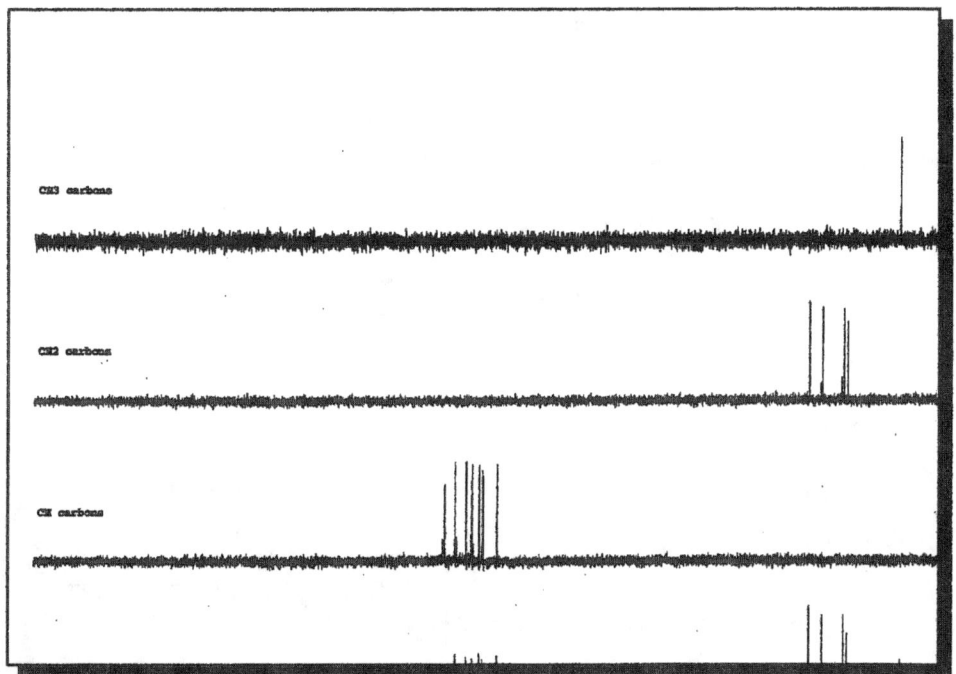

Figure 68 DEPT Spectrum of Ligand **68**

104

For the expected 21 different carbon signals, five should be in the alkyl region and sixteen in the aromatic region, which is seen in the spectrum. Also, of the sixteen aromatic signals, seven are from quaternary carbons and should be more suppressed more than the nine protonated carbons. Again, this expectation is seen in the ^{13}C NMR spectrum of ligand **68** (**Figure 67**). Similarly, the quaternary carbons should, and do, disappear from the DEPT spectrum (**Figure 68**). The DEPT spectrum also allows the certain assignment of C1 to the only methyl group of the ligand. C4 and C5 can be

assumed to correspond to the 1-position of the indenyls based on previous spectra. The C2 and C3 carbons must be the remaining methylenes of the bridge. To absolutely determine the assignments of the remaining carbons would require HMBC and HMQC spectra. Unfortunately, the sample kept for characterization was too small, after Prep TLC, to obtain these spectra.

However, the sample was large enough for mass spectrometry. The mass spectrum for ligand **68**

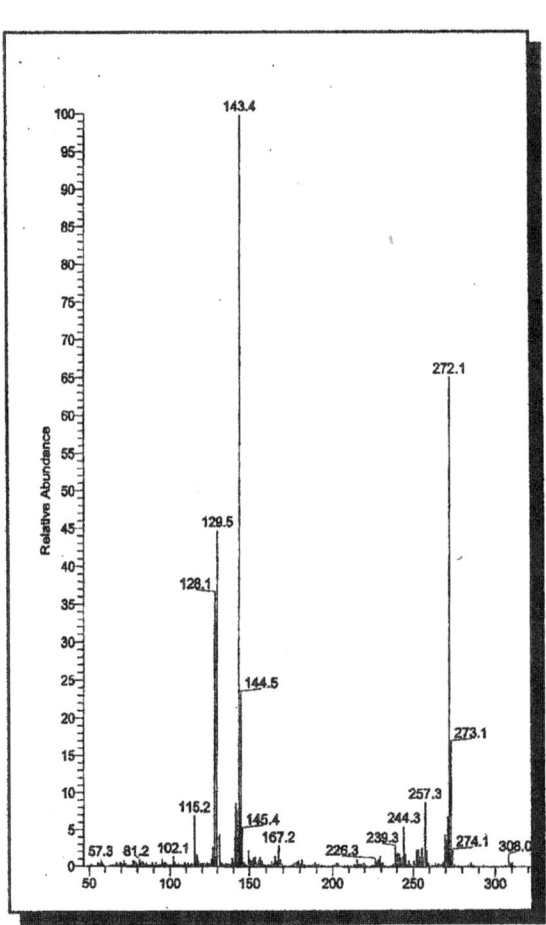

Figure 69 Mass Spectrum of **68**

105

(**Figure 69**) was obtained. The base peak corresponds to the 2-indenyl fragment with the ethylene bridge attached and the molecular ion is the next most abundant fragment. The unusually good behavior of ligand **68** in the mass spectrum was also worthy of note. The time profile, shown in **Figure 70**, shows a single large peak, corresponding to ionization that is sudden and complete, emphasizing the purity of ligand **68** and its

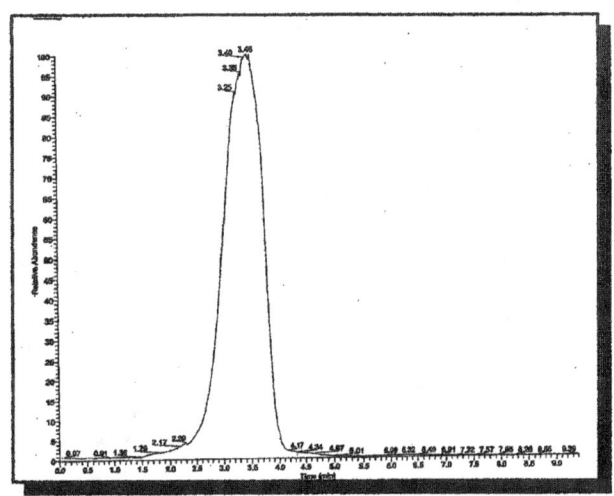

Figure 70 MS Time Profile for **68**

stability. Again, the HRMS was obtained via the + Na method. The mass found was 295.1446 while the calculated mass was 295.1463 for the formula $C_{21}H_{20}Na$.

In repeat runs of this procedure, the p-toluenesulfonyl chloride repeated its destruction of a majority of the compound, therefore the route was changed to mesylation in triethyl amine. This achieved the same bis-indenyl ligand in better yields. On the first attempt 22% of the product ligand and 37% of the starting material were recovered even after a purification column. The rest of the material appeared to be the mesylate.

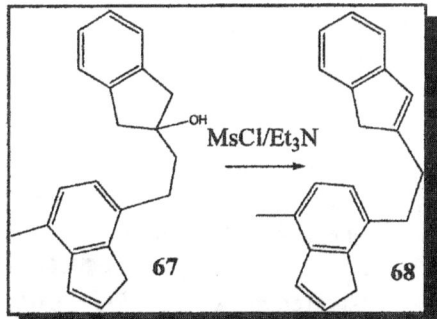

Figure 71 Mesylate Elimination

106

Since starting material was recovered and what was believed to be mesylate, on the second attempt the reaction was again allowed overnight to form the mesylate, but then was refluxed for 2 hours to complete the elimination. On this second run, 94% of product ligand and a trace of starting alcohol were recovered from the purification column.

These changes in the procedure made the synthesis reasonably efficient and the ligand, having been characterized enough for confirmation, was metalated by typical procedures. The ^{1}H NMR spectrum (**Figure 73**)

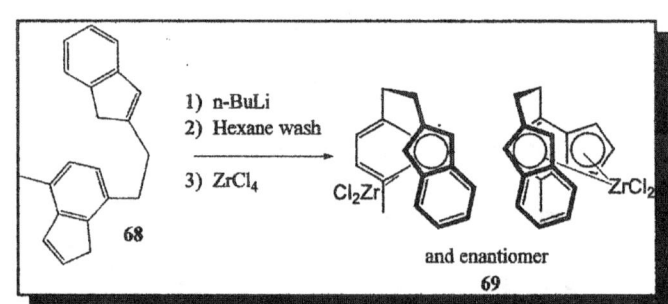

Figure 72 Metalation of Ligand **68**

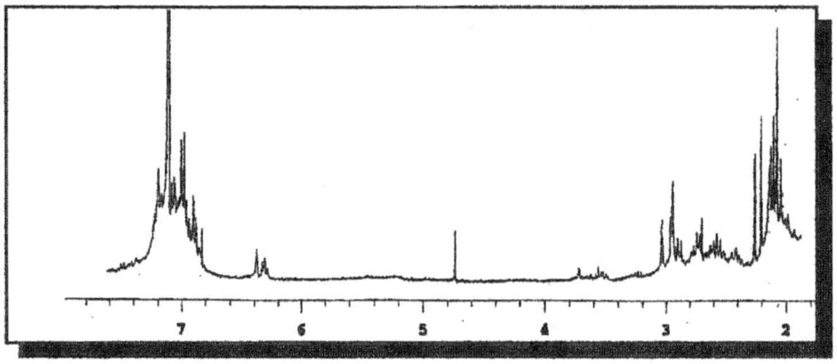

looked reasonable, but messy. A repeat of the metalation again gave metallocene **69**, though again it seemed quite messy. A ^{1}H NMR spectrum of the glovebox deuterated solvent showed many

Figure 73 ^{1}H NMR Spectrum of Metallocene **69**

impurities in the 0.5 δ to 2.1 δ region, explaining the business of the metallocene's

107

spectrum. However, since the methyl peaks of metallocene **69**, shown in **Figure 74**, are nearly free from the impurities, much can be learned. While it is possible that the two peaks correspond to the two torsional isomers of metallocene **69** (**Figure 75**), we expected the inter-conversion to be fast. It is possible that our expectation is wrong. As mentioned at the beginning of the chapter, the zirconacycle formed in this metallocene is a six-membered ring, rather than the usual five-membered ring. It is possible that

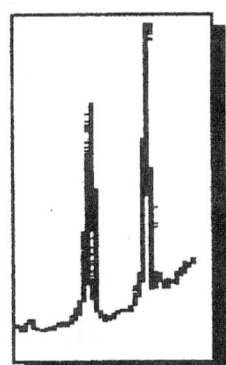

Figure 74
Zirconocene 69
and Dilithium
Salts

at room temperature the peaks to do not coalesce as expected. However, after making the models and noting the ease at which the two twist boats interconvert, it is more likely that the two peaks

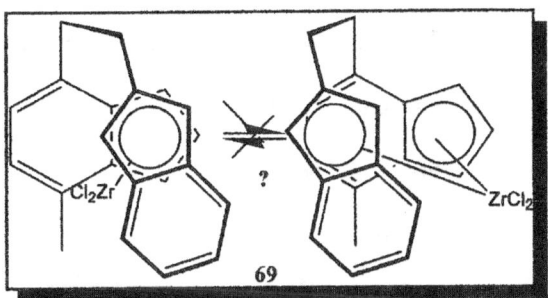

Figure 75 Torsional Isomers of Metallocene **69**

correspond to the lithium salt of the ligand and the metallocene. Assuming the two peaks correspond to the dilithium salt and the metallocene, the torsional isomers of the enantiomers are interconverting quickly on the NMR timescale.

Attempts to purify the zirconocene from the dilithium salt were unsuccessful as the salt had solubilities quite similar to the metallocene. However, a decent ^{1}H NMR spectrum (**Figure 76**) of the pair was obtained in deutero-methylene chloride after the tetrahydrofuran (THF) soluble material was filtered through a fritted filter (under argon).

108

Other than remaining
THF and some
hydrocarbon impurity
from the glovebox
benzene, the spectrum
was fairly clean,
allowing good enough
integration of the peaks
to generally assign

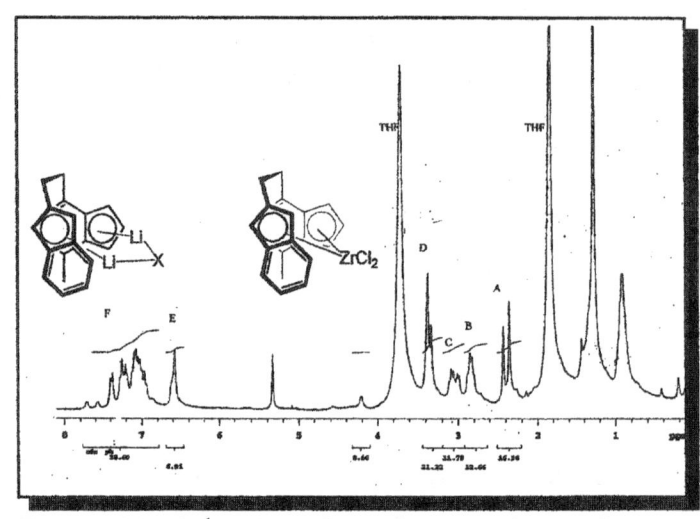

Figure 76 Crude ^{1}H NMR of Metallocene **69** in CD_2Cl_2

hydrogen sets in the spectrum to the hydrogens of the metallocene. The A hydrogens
integrate together to 3H and correspond
to the methyl group. The B and C
hydrogens each integrate together to be
2H and correspond to the methylenes of
the bridge. The D hydrogens together
integrate to 3.5 H and the E hydrogens
together integrate to 1.5 H. In general,
they correspond to the five hydrogens
of the two cyclopentadiene units (in
each compound). Similarly, the six
hydrogens or the aromatic groups are
grouped together as hydrogens F which

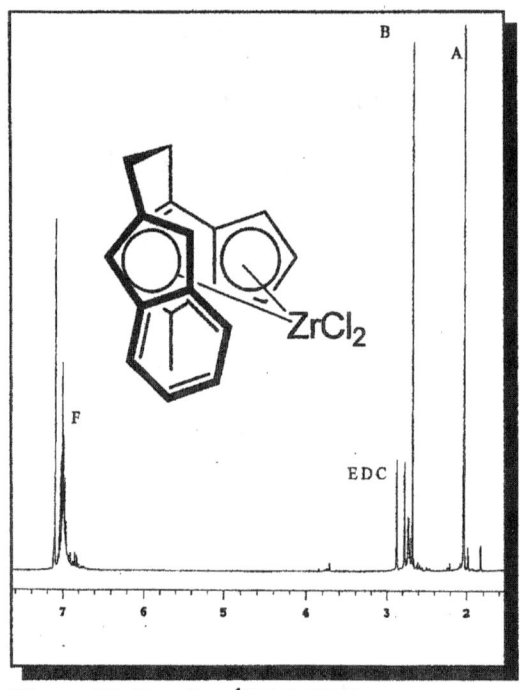

Figure 77 Puzzling ^{1}H NMR Spectrum

109

integrate to 6H together.

In order to avoid the problem of the separation, the reaction was run with a slight

excess of the zirconium tetrachloride. This time the ^{1}H NMR spectrum (**Figure 77**) was

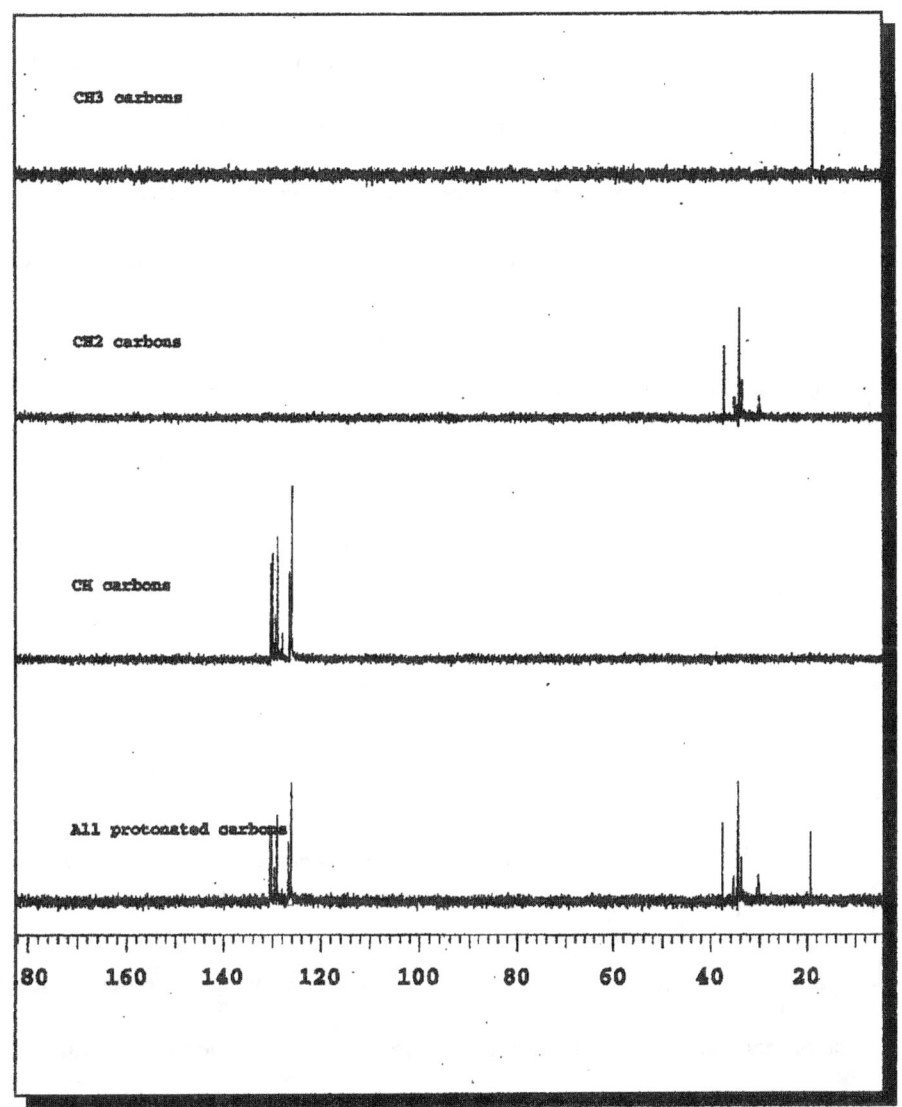

Figure 78 DEPT Spectrum of Metallocene **69**

110

very clean, but puzzling. All the aromatic hydrogens (hydrogen set F) appear in a conglomeration centered on 7.0 δ and the rest of peaks in the spectrum appear in the 2-3 δ region, A, B, and D as singlets. The integrations also left us somewhat confused. Hydrogen set A integrated to 6H, B to 3H, C to 2H, D to 1H, E to 1H, and F to 8H. Based on past experiences with the methyl group at just above 2 δ, we assumed that three of the hydrogens in the A set were from the methyl. By the chemical shift of the D and E hydrogens at almost 3 δ, they were assumed to be the two diastereotopic hydrogens of the bridging carbon nearest the inden-4-yl ring. Hydrogen set C, a multiplet integrating to

2H, was assumed to be the hydrogens on the other bridging carbon. The other three hydrogen of set A and the three hydrogen of set B remained puzzling while the 8 H of set F were assumed to be eight of the eleven aromatic indenyl hydrogens (counting the cyclopentadienyl hydrogens).

The DEPT spectrum (**Figure 78**)

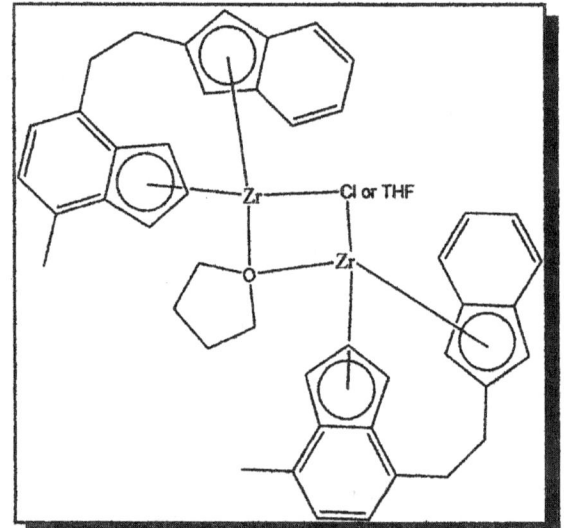

Figure 79 Hypothetical Zirconocene Dimer

was heartening, for it showed the expected: a single methyl, two methylenes of significant quantity, and a forest of methines in the aromatic region. It also showed four methylenes of lesser quantity. Taken with the extra three hydrogen in set A and the three hydrogen of set B, the four methylenes are suggestive of the possibility of

111

tetrahydrofuran. However, the peaks of tetrahydrofuran in the ^{1}H NMR spectrum are usually found at 3.58 δ and 1.78 δ. The ratios also account for only 3 molecules of tetrahydrofuran for each four molecules of product (the four hydrogen of THF integrating to only 3H). However, if the tetrahydrofuran were forming a bridging unit with the zirconocene, the oxygen's binding to the zirconium could result in the observed change in shift. To sum up, it is possible that the zirconocene is actually existing as some form of dimer with bridging via 1 or 2 tetrahydrofuran molecules. This would account for the DEPT spectrum looking as expected except four the four extra methylene units and the extra hydrogens in the ^{1}H NMR spectrum. The HMQC spectrum furthers the strength of this seemingly far-fetched idea. The aromatic methylenes at 7.0 δ appear to be on at least ten different carbons. Taking into account that aromatic hydrogens are often suppressed, the initial 8H integration must be incorrect and at least 10 H are actually found in that region. Perhaps counteracting ring currents in the other aromatic species can account for this anomaly.

This compound, which should be in a Π-conformation, should be submitted for polymerization tests. A comparison of its activity should give at least a general idea on the structure/activity relationship of ϒ-conformations, though like the Kaminsky Complex [166] the alkyl group at the 2-position (the bridge itself here) could be responsible for a moderate increase in activity. However, it should still be examined.

112

4.5   Routes towards the Synthesis of Ethylene-bridged-1-(7-Methylinden-4-yl)- 2-(inden-1-yl)zirconium(IV) dichloride

To this point, the phenyl-containing bridge had not reached ligand stage, let alone metallation. The propylene-bridged 7,1'-system had been too conformationally mobile to isolate a single isomer. The ethylene-bridged 7,2'-system was not crystallized, was racemic, and was substituted in the 2-position. However, with less conformational mobility than the propylene-bridged system, the ethylene bridged 7,1'- version should show a preponderance of the metallocene in the energetically preferred conformation. The meso-like isomer would, if formed, be expected to reside in a Γ-conformation while the rac-isomer should stay in a Υ-conformation. The 7,7'-ethylene version by Combs and Halterman had significant preference for the

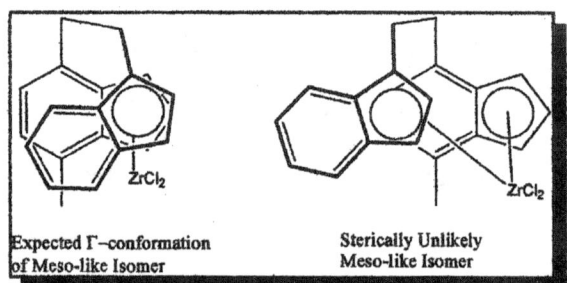

Figure 80 Expected Preference of Meso-like Isomer for a Single Conformer

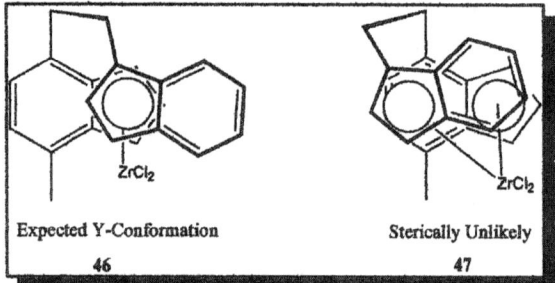

Figure 81 Expected Preference of Rac Isomer for a Single Conformer

rac-isomers over the meso-isomers. With this in mind, we expect to see at least a significant preference for the Υ-conformation of the rac-isomer and hope for its exclusive

113

formation.

Several different pathways were investigated towards the preparation of this

ligand and its zirconocene. The first route considered (**Figure 82**) was a Michael

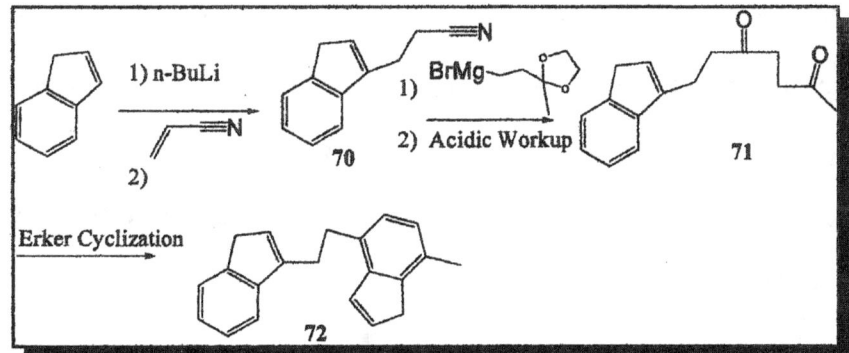

Figure 82 Originally Envisioned Route to 7,1'-Ethylene-bridged Ligand **72**

addition of lithium indenylate to the α ,β unsaturated propionitrile [132]. This nitrile

could then be attacked by Grignard **55** from the cinnamate route to form, after acid

hydrolysis and deprotection, diketone **71** that the Erker cyclization [100]could convert to

desired ligand **72**.

Unfortunately, the first reaction was quite touchy. The very first attempt appeared

to work well, giving a product with a distinct odor rather like rat poison. However,

attempts at purification used up the small stock before a clean ¹HNMR spectrum could be

obtained and many attempted repeats
ended up with polymeric messes.
Therefore, an alternate substrate, 3-
bromopropionitrile, was used for the

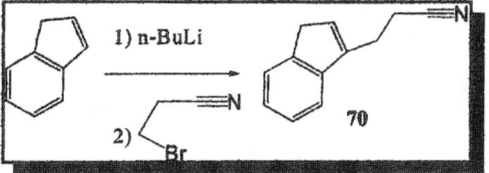

Figure 83 Formation of Nitrile **70**

114

indenylate to attack (**Figure 83**).

While the chemistry is essentially the same, the 3-bromopropionitrile was much more reliable, again forming a product with the (unenviable) odor of rat poison in 60% yield, after purification by silica gel column (and separation of the minor double-bond isomer). The ^{1}H NMR spectrum (**Figure 84**) shows the expected 1H:2H:1H peaks in aromatic region between 7 δ and 8 δ (E, F, G)and a 1H peak at 6.4 δ (D) and three 2H peaks in the 2.5-3.5 δ (A, B, C). Based on Silverstein's text [176], simple hydrogens on carbons near a nitrile group should show near 2.6 δ. Ours, being beta to an aromatic ring,

should be slightly higher and appears near 2.8 δ, the lowest shift hydrogen set and so assigned A. The typical 1-position indenyl peaks are found near 3.5 δ and here are found at 3.4 δ, C. The last methylene is near the

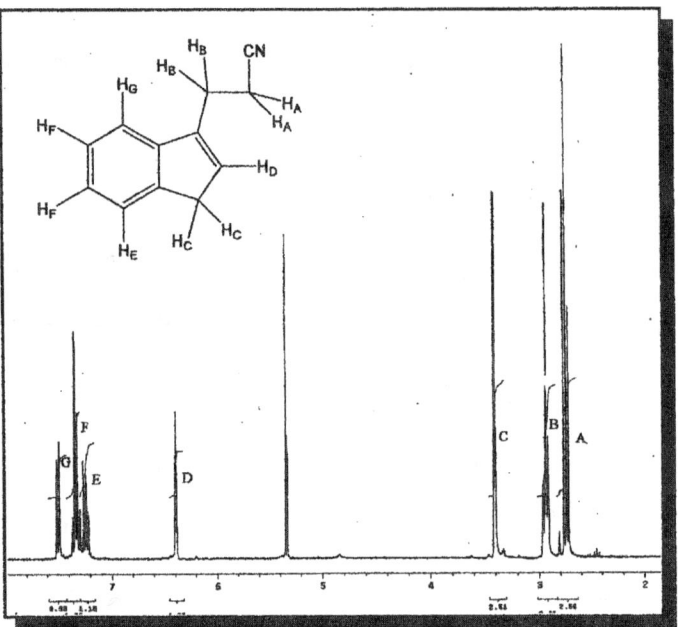

Figure 84 ^{1}H NMR Spectrum of Nitrile **70**

ring, and, as seen often in the past, shows at nearly 3.0 δ, B. Typical of indenyl aromatic hydrogens, the hydrogen of the carbon nearest the double bond is shifted higher, G, the one farthest away lower, E, and the two in the middle overlap, F.

115

The COSY spectrum (**Figure 85**) helps confirm these assignments, showing the

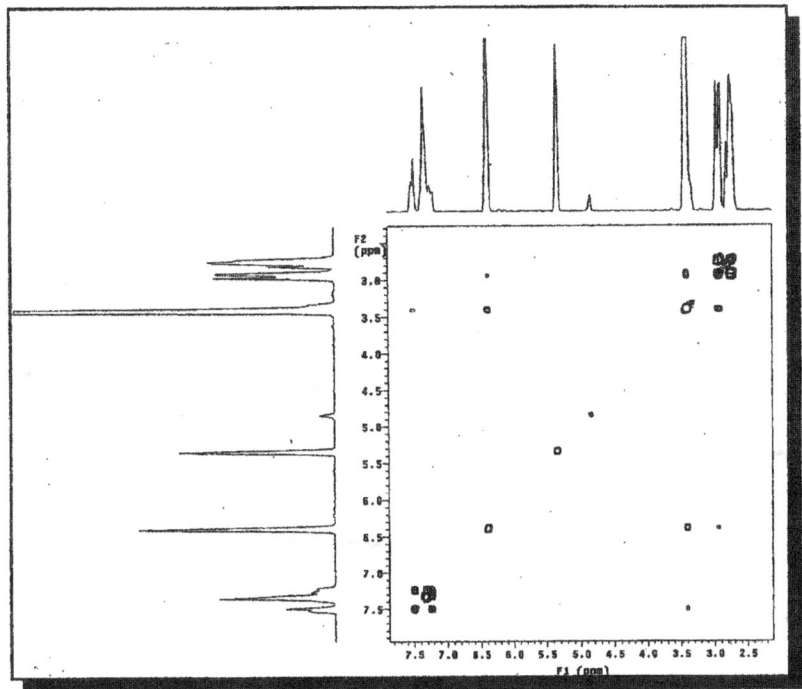

Figure 85 COSY Spectrum of Nitrile **70**

coupling of hydrogens A with B; B with A, C, and faintly with D; C with B, D, and

faintly with G; and, not

surprisingly, the aromatics

with themselves.

The ^{13}C spectrum

(**Figure 86**) shows eleven clear

signals and one possible one.

C1-3 were assigned based on

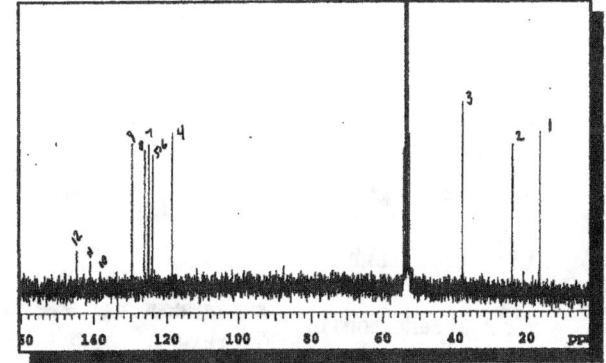

Figure 86 ^{13}C Spectrum of Nitrile **70**

116

previous experience with indenyls. C6 is assigned based on the fact that it is the lowest

quaternary carbon and the fact that aliphatic nitriles usually appear in the 117-120 ppm

range of a carbon spectrum [176]. C4, 5, 7, 8, 9 are strong signals characteristic of

protonated aromatic and alkenyl carbons while C10, 11, 12 are weak (suppressed) typical

of quaternary carbons. Since nitrile 70 has five aromatic or alkenyl methines and 3

quaternary carbons
(exclusive of the nitrile
itself), the structure is
consistent. The DEPT
spectrum (**Figure 87**)
confirms the methine and
quaternary assignments.

With nitrile 70
reasonably well
characterized, the Grignard
addition towards the
formation of
diketone 71 was
attempted (**Figure
88**). While similar
Grignard additions to
nitriles are known in

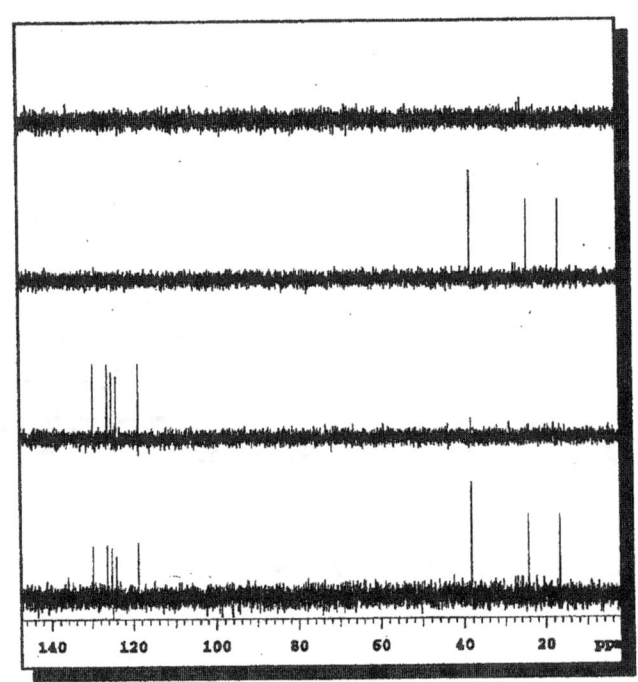

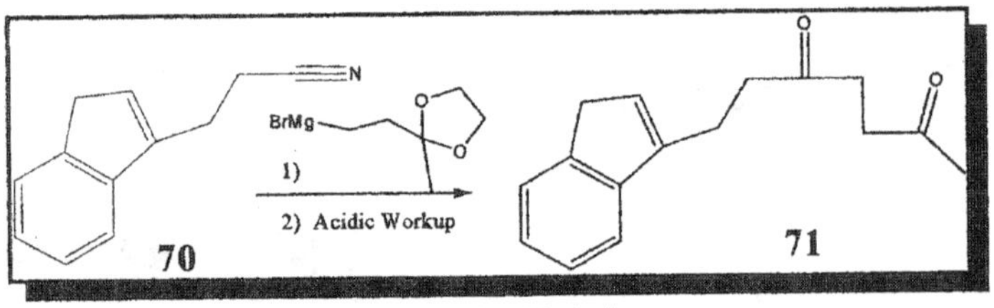

Figure 87 DEPT Spectrum of Nitrile **70**

Figure 88 Attempted Grignard addition

117

the literature [133], this one did not produce the desired diketone. After many fruitless

attempts, the Grignard route was abandoned.

Again, the reaction sequence was modified. Rather than attempting the Grignard

addition, it was decided to reduce the nitrile to the aldehyde and use the Stetter/Erker

procedure that had worked so well before. Since the indanoic nitrile was relatively facile

to make via 3-bromopropionitrile, a larger scale reaction was performed giving 8.6 grams

of the indanoic nitrile 70 (91% yield). With this large quantity of nitrile, several attempts

at converting the nitrile to aldehyde were attempted. The most successful of these was a

Raney Nickel

reduction [134], that

actually formed the

aldehyde in 82% yield.

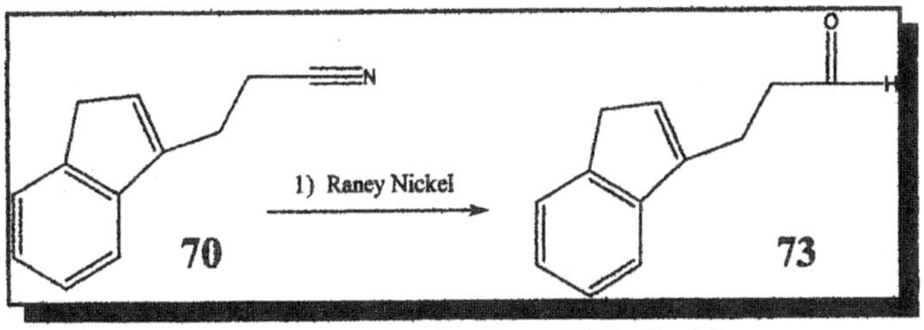

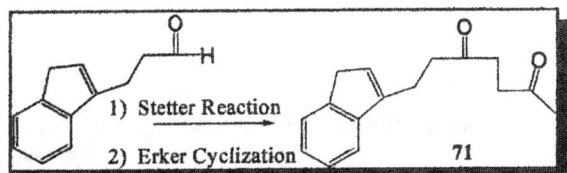

Figure 89 Raney Nickel Reduction of Nitrile 70

An attempt to

use the combined Stetter/Erker procedure on aldehyde 73 from the Raney Nickel

reduction resulted in diketone 71 (64%). Apparently the Stetter had worked, but not the

Erker. The ^{1}H NMR looked rather like a combination of the 4,7-dioxooctanoic acid

spectrum and the 7-methylinden-

4-ylpropanoic acid spectrums. A

mass spectrum showed the

molecular ion as a fairly

significant contributor. However,

Figure 90 Stetter/Erker of Aldehyde 73

there was not enough material to perform significant purification, and another route was

118

developed in the hopes of finding a higher yielding sequence. This route was then

abandoned.

Reflecting back to the previously successful routes to the propylene bridged 7-1'

and ethylene bridge 7,2'- ligands, an attempt was made to form this ligand via the same

route with one methylene unit less in the

starting diacid. Using oxalacetic acid

rather than the 2-ketoglutaric acid

seemed likely to be an exact match and

known chemistry could be used to

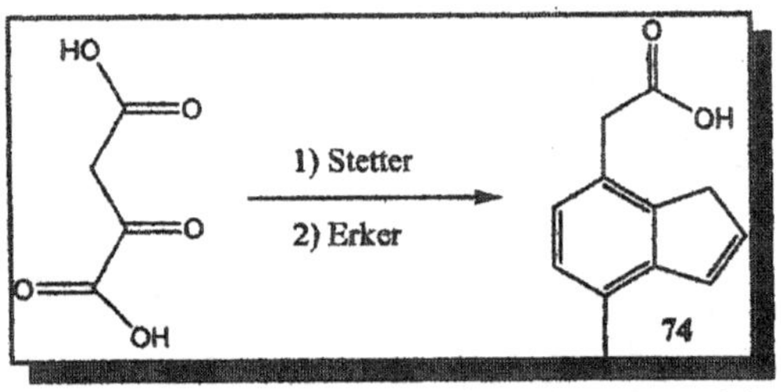

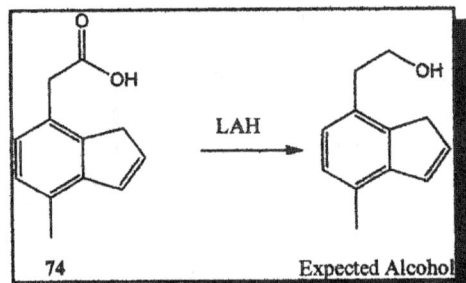

Figure 91 Stetter/Erker of Oxalacetic Acid

achieve the desired product. In fact, the

combined one-pot method of the Stetter

followed by the Erker worked well. A crude

^{1}H NMR spectrum showed complete loss of

starting acid and peaks consistent with the

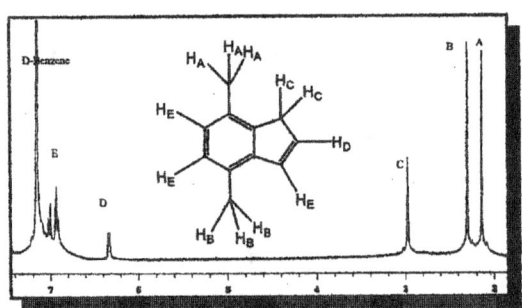

Figure 92 Attempted LAH Reduction

expected indanoic acid **74**. The crude

material, as usual, was subjected to

lithium aluminum hydride reduction.

However, unlike the previous sequence,

this time the lithium aluminum hydride

cleanly decarboxylated the compound

Figure 93 Decarboxylated **74**

to the well known 4,7-dimethylindene. An attempt to esterify indanoic acid **74** (in hopes

of preventing subsequent decarboxylation via lithium aluminum hydride) failed. It is

119

possible that the acidic nature of the hydrogens between the ring and the carbonyl allow for facile enol formation which prevented the esterification. However, whatever the reason, the esterification was unsuccessful. With three different routes not achieving the desired ligand, the ethylene-bridged route was placed on hold.

4.6   Conclusion

In summary, the initial pathway which started the investigations in the direction of a phenyl-containing chiral bridge remains an essentially good idea, but the quick success of the 'test' reaction made it the more judicious choice. The benefit of the initial route remains the possibility of including chirality in the bridging unit. This potential is lost in the successful routes. However, with the Stetter reaction so thoroughly studied, it should be relatively simple to run a Stetter on cinnamaldehyde. The Michael addition of the indenylate would then give a product only an Erker cyclization away from the initially desired ligand.

However, the 'test reactions' provided two new metallocenes and a route that includes the potential for many other metallocene structures by varying the anion which displaces the mesylate or the metalation procedure or even the metal itself. The first metallocene was formed in meso-like/rac diasteromeric ratios of roughly 60:40. Each diastereomer had only one peak in the room temperature ^{1}H NMR and thus was showing an average of its conformational isomers. Thus, this metallocene was determined to be too mobile in the bridging unit to achieve a single conformational isomer.

120

The second metallocene, that with the bridge tethered to the 2' position of the second indene, is somewhat less interesting as each face of the inden-2-yl moiety is equivalent and forms merely enantiomers when metalated. However, these enantiomers should be in a fairly rigid II-conformation. It should, therefore, be submitted for polymerization experiments. Unfortunately, the 2-position substitution is known to increase the activity of metallocenes, so increased activity could be due to electronic effects in the substitution rather than structural geometry of the conformation. However, as Kaminsky's Υ-conformation is also substituted in the 2-position, a comparison between the activities of these two compounds should yield some information as to the structure/activity relationship.

The third attempted metallocene has the best potential for definitively answering structure to activity questions for the Υ-conformer. Many routes to the ligand of this metallocene were attempted unsuccessfully. However, with thesis preparation and the discovery of the actual yield in the Raney nickel reduction of the nitrile, this route to the vaunted ethylene-bridged metallocene could be pursued by new students fairly facilely.

In conclusion, while a few niggling details remain to be completed before final publication of these results, two metallocene formations were successful. The NMR spectroscopic characterization has been completed on these metallocenes. With such success, other ligands could be formed by similar routes and techniques. For example, the 7,1'-propylene bridged version has as its final step the S_N2 displacement of mesylate by indenylate. This displacement could as easily be accomplished with substituted indenylates, or even with the sodium cyclopentadienyl or fluorenyl moieties. However,

121

perhaps the greatest asset of this research is the success of the combination Stetter/Erker reaction to take aldehydes or oxyacids to indenyl moieties. Essentially any aldehyde or oxyacid could now be used to form a substituent at the 4-position of an indene.

122

Chapter 5

Chiral Bis-Indenyls

5.1   Introduction

Having pursued the project of intermediate conformational isomers of bis-indenyl zirconocenes for the better part of five years, we decided it would be worthwhile to attempt projects more along the usual Group Halterman line: projects involving chiral bis-indenes. The metallocenes of the previous chapter were chiral, but their ligands were not. We now turned out attention to the synthesis of metallocenes whose ligands were chiral. As seen in Chapter 2, chirality can be built into ligands in a variety of ways. Our attempts included an attempt in which the bridging unit was the inherently chiral binaphthyl and an attempt in which a chiral substituent (menthyl) was added. Chiral ligands have different transition state energies for the approach of metals to a specific face of the ligand. Thus, with chiral ligands the potential exists for selectively forming the diastereomer with the lower energies or even a specific enantiomer of that diastereomer.

The first chiral bis-indene we considered was the binaphthyl-bridged bis-inden-7-yl ligand. Group Halterman has produced quite a few of these binaphthyl-bridged metallocenes [149]. However, the version with the indenes bound to the binaphthyl from

123

the 7-positions had yet to be achieved. Since some of the previous routes had passed through the known dialdehyde [151], the ease of the Stetter reaction on aldehydes seen in the last chapter suggested attempting the formation of tetra-ketone 76 and cyclizing it

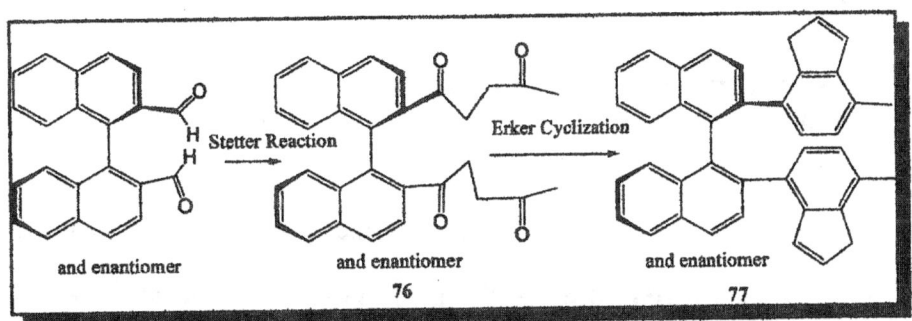

Figure 1 Theoretical Route to Binaphthyl-bridged 7,7'-ligand 76

with a double Erker Cyclization to ligand 77, more for the sake of completeness than any real desire for the ligand. However, while the 7,7' version will complement the known 1,1' and 2,2' versions, the methodology of a double Stetter reaction and double Erker cyclization will test the limits of these very useful reactions.

The other chiral ligand considered was the ethylene bridged-bis-2-menthylindene. In this case, the ligand is chiral due to the presence of a chiral substituent. While the unbridged version of this ligand has been used in metallocene formation [138], the bridged version has previously eluded synthesis. We believed the known palladium-coupling procedure between aryl halides and Grignard reagents [139] could rectify this lapse. By taking the known 1,2-bis(inden-1-yl) ethane [140] and brominating it in the presence of triethylamine the 1,2-bis(2-bromoinden-1-yl) ethane could be formed [141]. Along with the known menthyl Grignard [142], this aryl dibromide should form the

124

ethylene bridged bis-2-menthyl ligand in very short order (**Figure 2**).

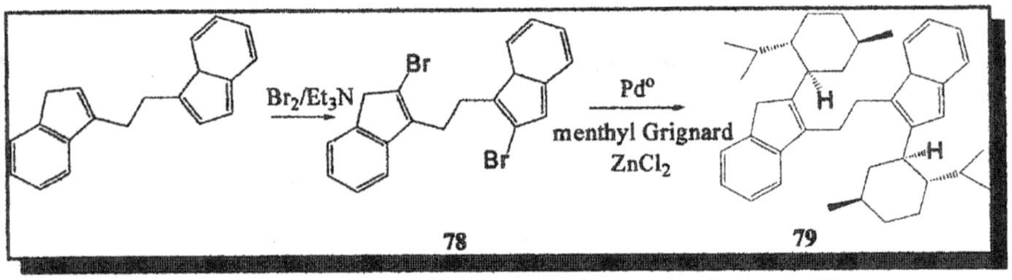

Figure 2 Expected Route to 1,2-bis(2-menthylinden-1-yl)ethane **79**

This bis(2-menthylindenyl)ligand, with the excessive bulk in the 2-positions should not form the meso-like metallocene on metalation. Additionally, the menthyl groups should direct the incoming metallocene to one face of each indenyl. Together the affects should cause the selective formation of the rac-diastereomer and possible even form one enantiomer of the rac-diastereomer selectively over the other.

These two chiral ligands were thus attempted. The first was attempted primarily for the sake of completion of the binaphthyl-bridged possibilities and to test the limits of the Stetter/Erker Reactions. The second was attempted for its potential to form rac-metallocenes, possibly enantioselectively, and because its non-bridged version was such an active metallocene [138].

5.2 Attempted Binaphthyl-bridged bis(inden-7-yl)Ligand

The first chiral ligand to be examined is the binaphthyl-bridged ligand **77**. While Halterman is already well-known for binaphthyl-bridged ligands [149], the version

125

tethering the 7-position of one indene to the 7-position of another has not yet been performed. Our previously established methodology of using the Stetter Reaction and Erker Cyclization to convert aldehydes and oxyacids to indenyl units suggested a short route to fill in this gap. In fact, from the known binaphthyl dialdehyde the desired ligand is only a double Stetter and double Erker away.

The known binaphthyl dialdehyde was synthesized from a sequence of known reactions starting with a Grignard self-coupling procedure [153] performed in the

presence of nickel dichloride (1.5 mole %) and triphenylphosphine (3.0 mole %). With this procedure, 2-bromo,1-methyl-naphthalene was coupled to form the 2,2'-dimethyl-1,1'-binaphthyl in good yield (56% compared to the literature

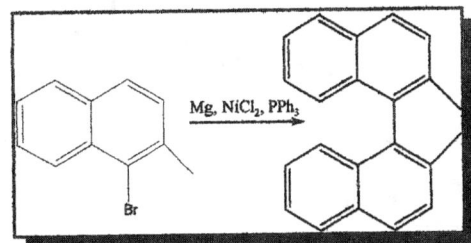

Figure 3 Grignard Self Coupling [153, 95]

value of 61% [153, 95]). NMR spectra matched the literary values and a mass spectrum found the molecular ion peak as the base peak.

Next, the dibromide was produced according to the procedure of David Combs [154] from excess N-bromosuccinimide (NBS) and light. The NBS and light subjugation was immediately repeated on the crude dibromide material to produce the tetrabromide in good yield (53-57% from the

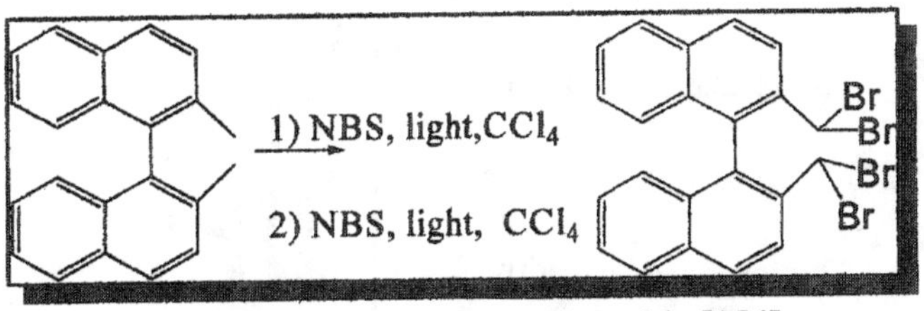

1) NBS, light,CCl$_4$

2) NBS, light, CCl$_4$

Figure 4 Formation of Known Tetrabromide [154]

126

binaphthyl). Again NMR spectra matched the literature values [154], the most significant of which are a 2H singlet appearing in the dibromide and disappearing in the tetrabromide at 4.2 δ from the first bromine on each methyl and the final appearance of a 1H singlet at 6.2 δ from the second bromine addition. In subsequent runs, NBS and light were reapplied to incomplete reactions until all starting material and all dibromide had been consumed. An initial mass spectrum was rather disappointing in the molecular ion's inability to extend beyond the background noise (**Figure 5**). However, persistent effort on the part of the mass spectroscopist finally achieved a run in which the molecular ion and the first three subsequent losses of bromine can be readily detected (**Figure 6**).

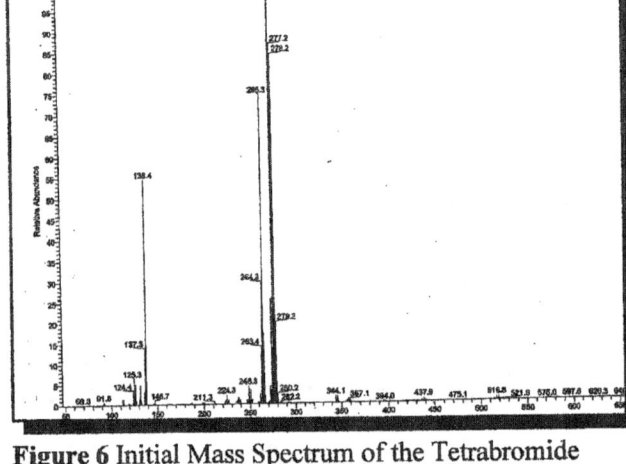

Figure 6 Initial Mass Spectrum of the Tetrabromide

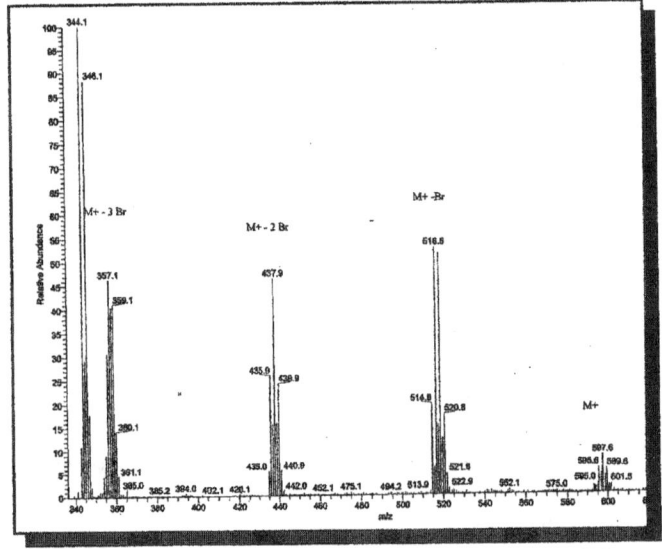

Figure 5 Best Mass Spectrum Expanded from 340-620

127

This tetrabromide was then converted by known methodology [155] to the aldehyde with silver nitrate in ethanol and water to form the

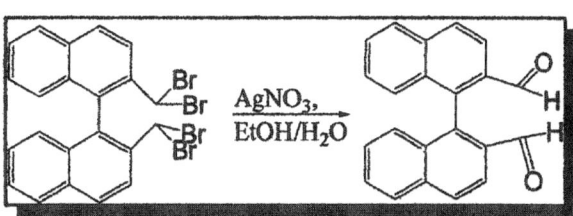

Figure 7 Oxidation to Aldehyde

dialdehyde in essentially quantitative yield (99.8% after recrystallization). On subsequent runs, over-oxidation to acid was occasionally a problem, sometimes cutting the yield to as little as 65%. However, though problematic for the yield, a simple silica plug conveniently removed the acid when over-oxidation occurred.

This dialdehyde was initially subjected to the combined, single pot procedure of the Stetter/Erker reactions worked out in the 7-1' and 7-2' bis-indenyl ligand systems. However, the product, when analyzed by ¹H NMR spectroscopy, showed no starting aldehyde, but also no 3H

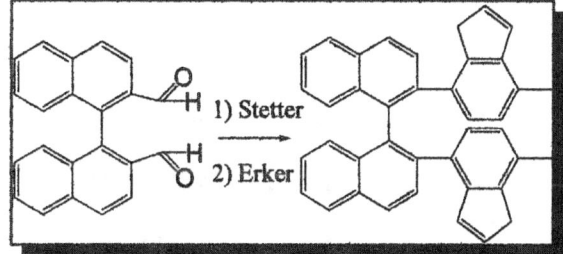

Figure 8 Attempted Combined Stetter/Erker to **77**

singlets corresponding to the expected ligand's methyls. There were, however, traces of material that seemed reasonable for the tetra-ketone were present, so the Erker Cyclization was repeated on this crude material. After two days reflux in sodium methoxide/methanol and freshly cracked cyclopentadiene an aliquot showed no sign of ketone peaks in the ¹H NMR spectrum (**Figure 9**) as well as peaks in all the regions expected for the bis-indenyl product, especially the characteristic peaks of the aromatic

128

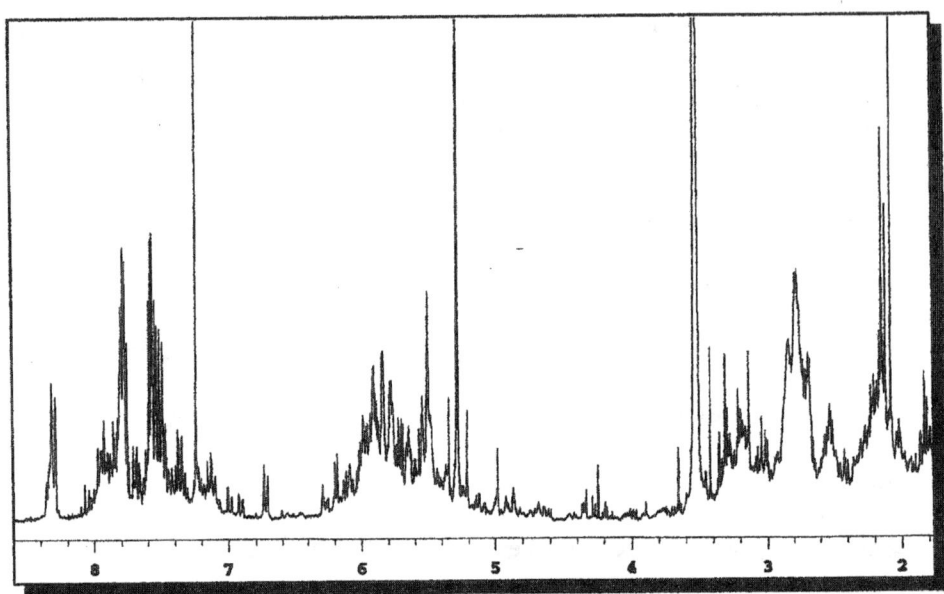

Figure 9 Crude Binaphthyl Ligand **77**

peaks at 7.0 δ, the alkenyls at 6.8 δ and the methyls around 2.1 δ. There were also four

such methyls, which would correspond to the four double bond isomers of the ligand.

After typical work up, the remnants were extracted with petroleum ether to produce a

beautiful golden solution that concentrated to an oily reddish residue containing

cyclopentadiene dimer and peaks quite possible for the expected product, but also

impurities. The residue was dissolved in methylene chloride and washed with water.

After the methylene chloride was removed, the compound was plugged through silica gel.

Unfortunately, the "impurities" were still there. Therefore, a long column was run

with straight petroleum ether. The first fraction containing UV-active material appeared

the most likely when analyzed by ^{1}H NMR spectroscopy. It contained material that

looked very reasonable for product. Other fractions contained peaks reminiscent of ethers

and alkenes. If indeed the first compound was product, it was formed in a little less than

129

4% yield, counting the remaining cyclopentadiene dimer that was unable to be removed from the product. However, the tiny sample was submitted for mass spectrum analysis.

While the molecular ion was not present, the base peak represented a C_7H_7 (tropylium-like?) fragment not seen in the other binaphthyl compounds and the 4-methylindenyl mass of 129 was nearly the size of the base peak. This rather suggests that the 4-methylindenyl

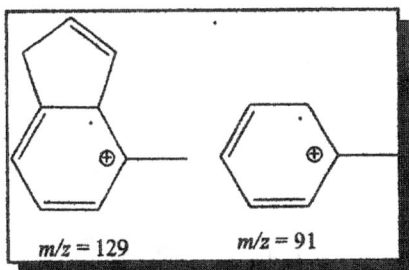

$m/z = 129$ $m/z = 91$

Figure 10 Major MS Fragments

component is giving rise to the base peak and that product is in fact present. However, since this route and compound were being pursued primarily for the sake of completeness, it was decided not to pursue the minuscule amount of material (122 mg, which included significant amounts of cyclopentadiene dimer, from a 5g scale reaction). Suffice it to say, if the product was formed, it was in such very poor yield as to be worthless as a synthetic step.

A repeat of the Stetter Reaction on the dialdehyde was performed for a longer period of time. This time rather than taking crude material on to the Erker cyclization, the tetraketone was isolated and purified. After purification, only a tiny amount of tetraketone (15 mg from 3 g of starting aldehyde) remained for a yield of less than ½ %. Obviously, these reactions do not do well on the double additions we were asking of them here. There are, apparently, limits to the utility of the Stetter reaction and Erker cyclization and attempts to do them on more than one position in the molecule at once exceeded these limits.

130

5.3 1,2-bis-(2-Menthylinden-1-yl) Ethane via Palladium Coupling

The next ligand system to be discussed is that of the ethylene-bridged bis-2-menthylindene. The proposed synthetic pathway (**Figure 11**) to this ligand is quite short

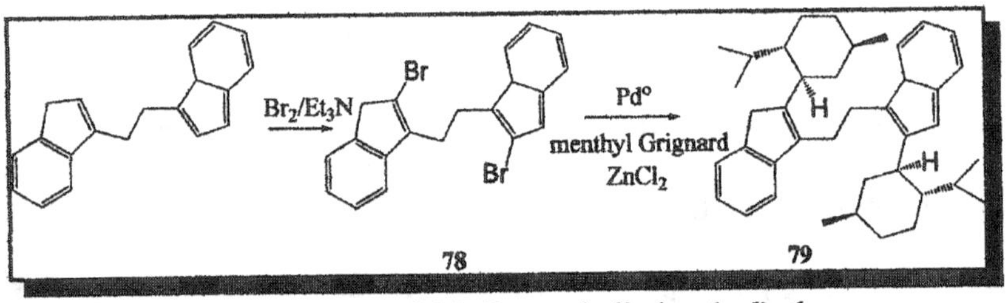

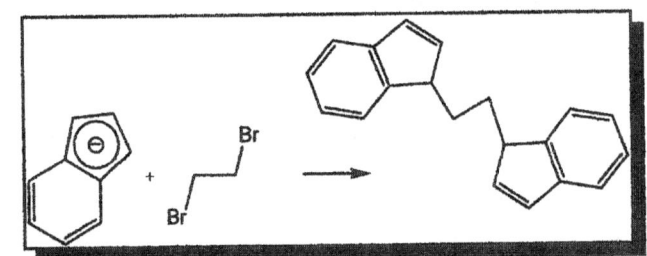

Figure 11 Pathway to *ansa*-1,2-bis(2-menthylinden-1-yl)ethane

and involves only one step not appearing directly in the literature, though the unbridged version and its metallocene do [144].

The first steps in the synthetic pathway are quite well known and were fairly simply accomplished. The starting bis-indenyl ligand was formed from the lithium indenylate addition to 1,2-dibromoethane exactly according to

Figure 12 Formation of the Starting Ligand

literature [140] in 32% yield on the first attempt. While literature claims a yield of 60% [140], this was never achieved. However, the reaction was run on such large scale that, even at 32% yield, enough of the product was formed for many bromination reactions.

131

Therefore, the reaction was not optimized. In fact, the reaction was only performed a few times. The various NMR spectra matched that in the known literature [140].

Similarly, the combined bromination/elimination (**Figure 13**) was run on large scale. Thus, its relatively low yields (50-60%) produced significant amounts of material for the attempted palladium couplings. Again, no attempt was made to optimize the reaction sequence since there was plenty of material with which to work. Again, the various NMR spectra matched the known spectra. However, some excellent crystals of bis(bromo) **78** prompted an x-ray crystallographic determination which had hitherto not been performed. The expected anti-orientation down the ethane and the rings pointing in opposite directions was heartening. These are the effects in the

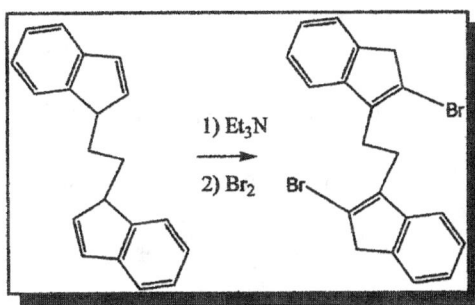

Figure 13 Bromination/Elimination

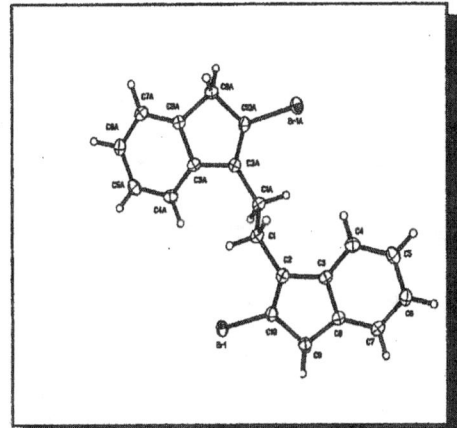

Figure 14 X-Ray Structure of **78**

proposed ligand expected to encourage rac-isomer formation. To see the much less bulky bis(bromo) **78** having the same effect boded well for the actual ligand.

It might be noted that bis(bromo) **78** retains its crystallinity even on the benchtop over months, as opposed to the 2-bromoindene which deliquesces in fairly short order if

132

not kept refrigerated and air-free. While the electronic effect of substituting an alkyl group at the 1-position should have destabilized the bromide, this was not the case. Perhaps the relatively rigid anti-orientation allows for better crystal packing such that the electronic destabilizing effect is overcome by the energy gained in the lattice structure.

When significant stockpiles of the *ansa*-1,2-bis(2-bromoinden-1yl)ethane had been amassed, the palladium coupling reaction was ready to be attempted. While Jason Shipman spent a lot of time and effort optimizing similar palladium coupling reactions [143] of brominated indenes to alkyl groups, the palladium tetrakis-triphenylphosphine that had served him so well did not work in significant yields to isolate any product. Nor did the

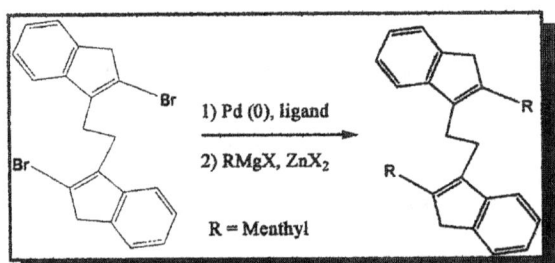

Figure 15 Palladium Coupling

palladium bis(ddppf) dichoride or the similar 1,2-bis(diphenylphosphino-) ethane nickel dichloride. Repeated coupling attempts with each of these catalytic systems failed at room temperature, in refluxing hexane, tetrahydrofuran, benzene, and toluene.

Following the advice of Dr. Vladimir Marin, a simultaneous attempt was made at coupling the cyclohexyl Grignard and the menthyl Grignard in side by side flasks to ascertain that the catalysts were truly active. The single attempt at this coupling with the cyclohexyl Grignard analog with the palladium bis(ddppf) dichloride showed coupled product with ^{1}H NMR spectra corresponding to the authentic sample provided by Dr. Marin. The molecular ion was also a significant contributor in the mass spectrum. Thus,

133

the conditions were correct, but the menthyl coupling was being especially stubborn.

Finally, the palladium tetrakis-triphenylphosphine coupling resulted in a smattering of product that was confirmed by mass spectroscopy. Unfortunately, the yield was so low that the sample could not be purified from the excessive amounts of menthyl Grignard byproducts and still retain enough material for even a good proton ^{1}H NMR spectrum. Attempted scale ups were similarly disappointing and other routes to this ligand were pursued.

However, the other routes that were attempted towards this 1,2-bis(2-menthylinden-1-yl) ethane, also ran into problems requiring significant synthetic effort to surmount. Therefore, the palladium coupling route, as the most direct route, was revisited and several more catalytic systems were tried.

Among these new systems were a couple in which the catalytic species were generated in situ. These two systems were formed from the tris-(dibenzylideneacteone) dipalladium at 5 mole percent with tri-o-tolylphosphine at 12 mole percent or 2-(dicyclohexylphosphino-)biphenyl at 12 mole percent. While the reactions were run exactly as before, the work-up was altered to evade the issue of purifying from the Grignard by-products that had haunted earlier efforts. The crude reaction mixture, kept in an aluminum-foil covered flask, was subjected to positive argon flow until all solvent (tetrahydrofuran) was removed. The residue was then triturated with hexane from the solvent still and injected while still under argon. Trituration under a positive argon flow resulted in a nicely yellow solution that was carefully removed to an argon-flushed round bottom flask. Unreacted menthyl Grignard and salts were left in the original reaction

134

flask. Once all unreacted Grignard had precipitated from the hexane, the hexane was

cannulated to another round bottom flask and removed from the product by roto-

evaporation to afford the *ansa*-1,2-bis(2-menthylinden-1-yl)ethane in reasonable yields

(40 % with tri-o-tolylphosphine and 21 % with 2-(dicyclohexylphosphino-)biphenyl).

Both of these *in situ* catalytic systems gave product in decent enough yield to

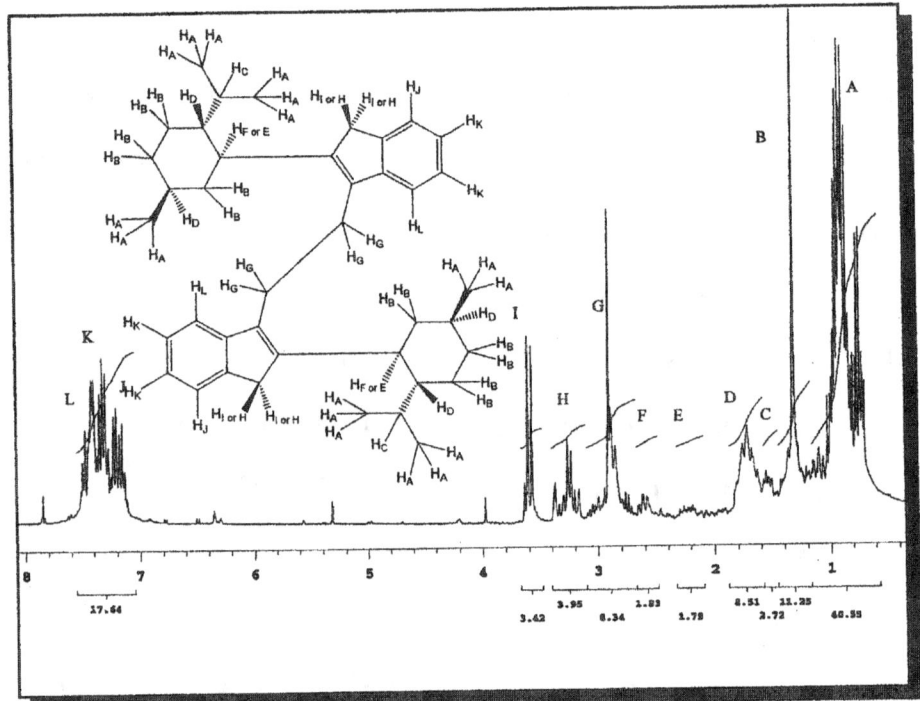

Figure 16 ¹H NMR Spectrum of *ansa*-1,2-bis(2-Menthylinden-1yl)ethane

characterize the ligand by ¹H NMR spectroscopy and mass spectroscopy. The ¹H NMR

spectrum seen in **Figure 16** can be used to roughly assign the hydrogens to the structure.

Overlapping hydrogen sets and the fact that every hydrogen is diastereotopic make

absolute assignments extremely difficult. However, the basic assignments are roughly as

135

given in **Figure 16**. Additionally, a slight deficit in the integration of the menthyl hydrogen sets and the presence of a peak at 6.25 δ, characteristic of hydrogens in the 2-position on indenyl units, suggests that some amount of the mono-menthyl compound has been formed as well. Based on the integrations, 66% of the menthyl peaks are from the bis-menthyl and 33% from the mono-menthyl.

The DEPT spectrum seems to bear this calculation out. The expected peaks (3 alkyl methyls, 3 alkyl methylenes, 2 allylic methylenes, 4 alkyl methine, 4 aromatic methine) predominate, but a significantly smaller amount of material exists in fairly similar patterns but with twice the number

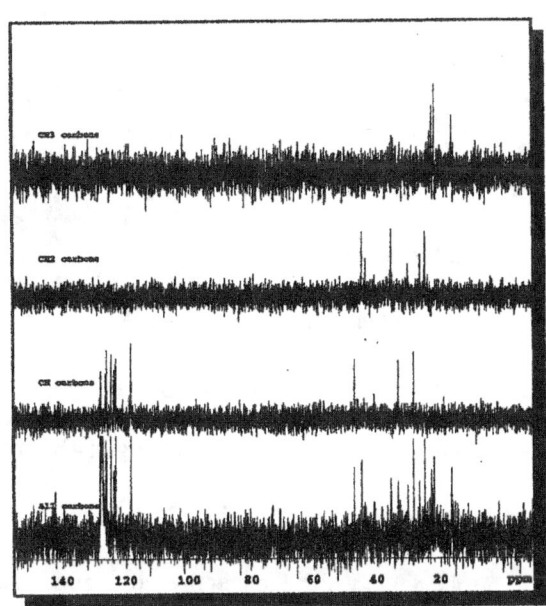

Figure 17 DEPT Spectrum of Ligand **79**

of aromatic methines. While integration may not be reliable in carbon-based spectra, it does give a general indication.

Since all the expected impurities are derivatives of Ligand **79**, the mass spectrum was not expected to be too useful except to confirm the presence of the molecular ion. In fact, though, the mass spectrum (**Figure 18**) also shows the presence of some of the bis(bromo)-**78** and the mono-bromide, mono-menthyl compounds. However, the molecular ion peak is present and the splitting between the bridging carbons is, next to

136

the menthyl base peak, the largest peak in the spectrum. Thus, while some impurities do exist, the predominance of the material is the desired ligand. Repeated attempts at purification were relatively unsuccessful.

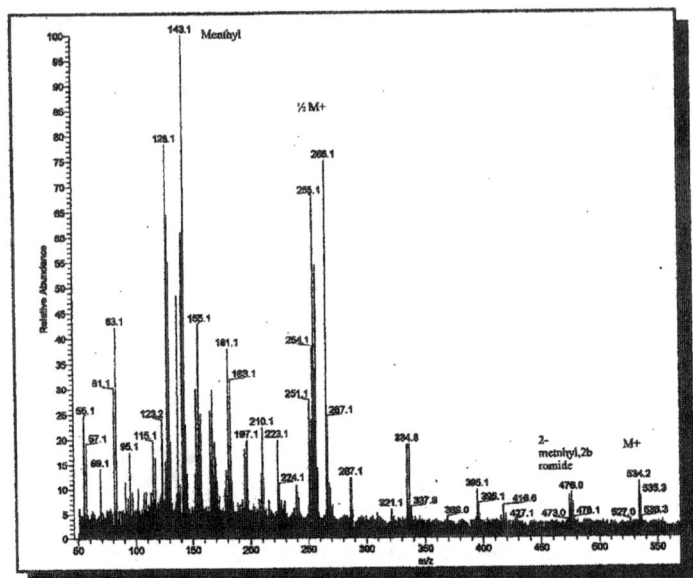

Figure 18 Mass Spectrum of bis(2-Menthyl)ligand **79**

Even excessively long silica gel columns with straight petroleum ether eluent could not completely separate the remaining bis(bromo) **78** from the bis(menthyl) product. Of all the attempts at purification, the best result was from recrystallizing the compound **78** in toluene, in which the menthyl compound is quite soluble. Repeated recrystallizations finally achieved fairly pure material, but at a great cost in yield.

With the difficulty in separating these two compounds, a much larger scale of the reaction was run for a much longer time (five days instead of the usual two or three) to hopefully convert all of compound **78** to product **79**. The ^{1}H NMR spectrum of this large scale reaction looked even better than previous runs, so it was subjected to a large excess of n-butyllithium and the insoluble dilithium salts washed with hexane and vacuumed to dryness.

The metalation of the dilithium salt with the zirconocene was effected by simply

137

massing the required amounts out of dilithium salts and zirconocene tetrachloride into an

argon-filled round bottom flask with a stir bar in the glovebox. Once the salts were

massed, the flask was moved to the Schlenk line, and toluene trickled in via syringe

through the septum while the flask was submerged in a dry ice/acetone bath under

positive argon pressure. After two days at room temperature, the toluene soluble

materials were cannulated from the remaining salts, evaporated under high vacuum and

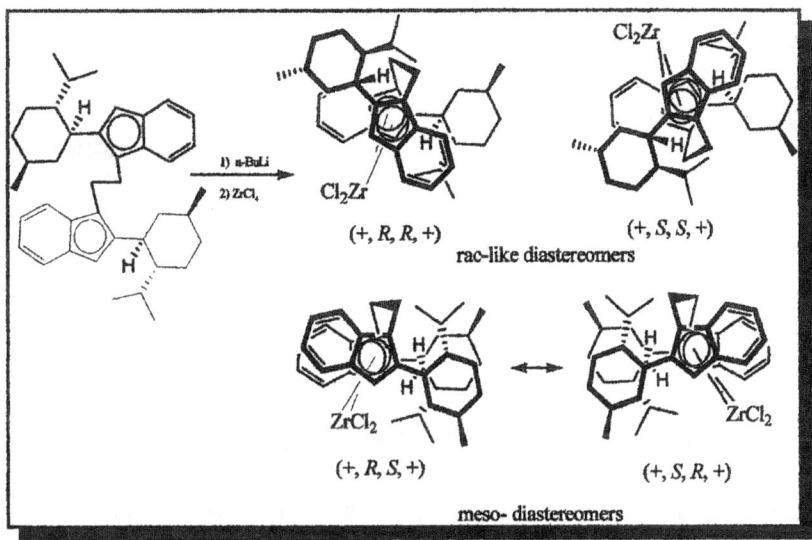

Figure 19 Metalation of bis(2-Menthyl)ligand **79**

examined by ^{1}H NMR spectroscopy.

Since the menthyl units are derived from a pure enantiomer [(+)-menthol], the

various stereocenters are set throughout the menthyl unit and are the same on both

indenes. The result is that each menthyl acts essentially as a single chiral center. By

denoting the entire menthyl unit as (+), the two centers of planar chirality derived from

metalation is made clear, as seen in **Figure 19**. Thus, changing each of the faces for the

138

two rac-like stereoisomers generates a different diastereomer (+, *R, R,* +) and (+, *S, S,* +) while changing both of the the meso-diastereomer faces gives the same (+, R, S, +).

The metalation worked surprisingly well considering the steric bulk of the menthyl groups. The ^{1}H NMR spectrum of the first metalation attempt showed near complete formation of metallocene, based on the absence of the ligand's indenyl peaks and the appearance of three obvious 1H peaks interlaced amidst the d-methylene

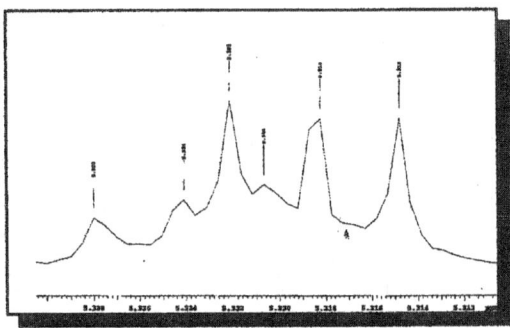

Figure 20 CD$_2$Cl$_2$ of Metallocene **80**

chloride peaks (used as ^{1}H NMR solvent for ease of integration) corresponding to hydrogens on the metalated cyclopentadienyl anion moiety of the indene. Changing solvents to deutero-benzene revealed another peak and caused them to shift to 6.0 δ, 6.2 δ, 6.4 δ and 6.6 δ. The presence of four sets of peaks for the 3-position proton suggests that both the C1-meso- and both C2-rac-diastereomers were formed. Attempts to separate the meso- and rac- metallocenes by solubility is occasionally successful in the literature. Unfortunately, the oily residue of this zirconocene could not be recrystallized as it was soluble in everything tried, even hexane.

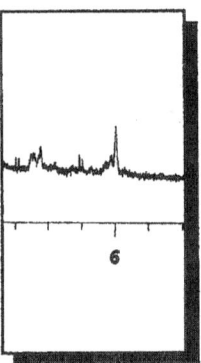

Figure 21
Zirconocene **80** in
C$_6$D$_6$

Hoping for a more selective metalation, Jordan's Method of metalation with

139

zirconium tetraamide was tried (**Figure 22**). The zirconium diamide version did allow the peaks previously hidden by the deutero-methylene chloride to be readily viewed, but there were still four of them.

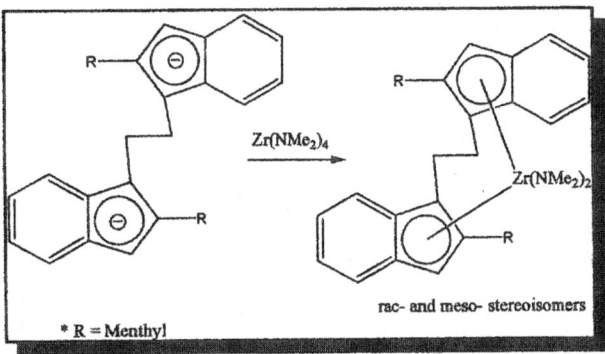

Figure 22 First Step in the Jordan Method

Typically, the diamide is converted to the dichloride after the fairly exclusive formation of rac-stereoisomers. However, as we still found both isomers, the zirconocene was abandoned. This unhappy end to the project was not a good note on which to end, so the titanocene was attempted.

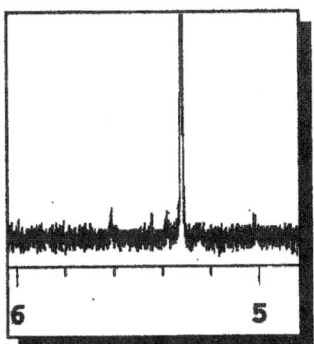

Figure 23 Results of the Jordan Method

The titanocene was formed in essentially the same manner as was the zirconocene, from the dilithium salt of the ligand and the metal halide. However, after the initial formation of the metallocene, it is acidified and oxidized.

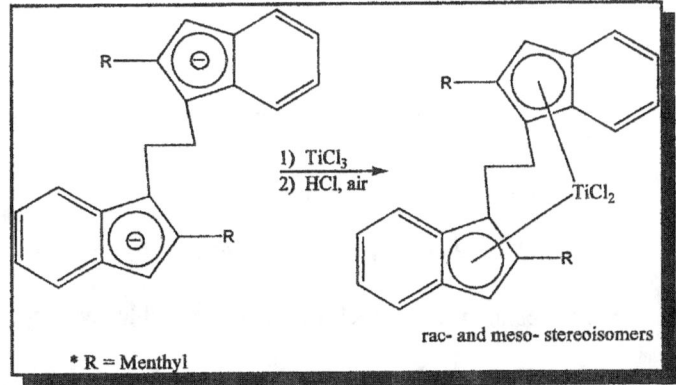

Figure 24 TiCl₃ Metalation of Ligand **79**

140

The procedure of Chen and Halterman for titanocene formation [152] was followed in a

slightly modified version. After the first day at room temperature, the reaction mixture was raised to approximately 60 °C for the second day. After the day at elevated temperature, chloroform and hydrochloric acid

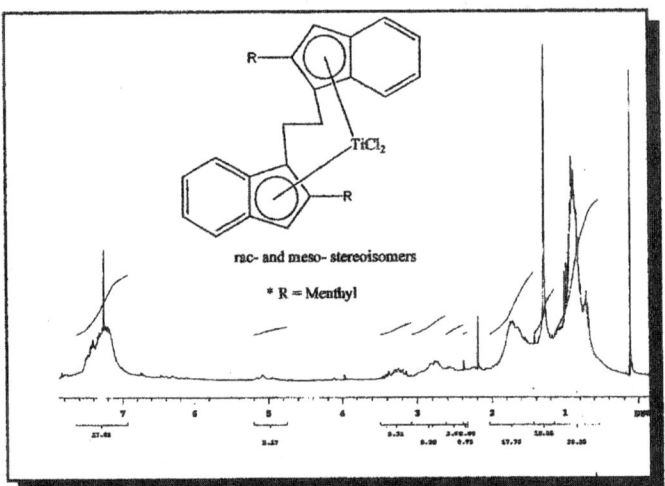

rac- and meso- stereoisomers

* R = Menthyl

Figure 25 ^{1}H NMR spectrum Titanocene version of **80**

(6M) in amounts equal to the original volume of toluene were added by syringe.

Compressed air was then bubbled through the solution for twenty minutes. Solvent was

removed to afford the crude reddish-orange titanocene as an oily residue. The ^{1}H NMR

spectrum was quite similar to the zirconocene analog, particular in regards to the location

of the cyclopentadienyl peaks. However, as the zirconocene, the titanocene showed the

presence of both rac- and the meso-like isomers, based on the number of these peaks.

Again, like the zirconocene, the titanocene was essentially soluble in everything.

Attempts to recrystallize from hexane afforded purer titanocene in the hexane solution

than in the precipitate. When the hexane was removed, an oily foam resulted.

An attempt to purify the titanocene by column chromatography resulted in a

yellow band moving quickly in methylene chloride that appears (by ^{1}H NMR spectrum)

after solvent removal to be starting ligand and a bright red band that does not move in

141

methylene chloride, which is assumed to be the titanocene. However, when diethyl ether was used to move the red band down the column, it ran quite nicely, but also changed to a golden yellow. Hoping that the golden yellow was a dilution effect didn't change the ^{1}H NMR spectrum (after solvent removal) that showed only traces of the expected titanocene peaks and some other indene-like peaks that could correspond to one of the indenyl ligands coming free from the titanocene.

In short, the bis-(2-menthylinden-1-yl) titanocene (IV) dichloride did not behave much better than the corresponding zirconocene. More attempts at column separation could still be attempted with different, non-coordinating solvents. However, both diastereomers are present and the expected selectivity not shown. Therefore, interest in the compound has significantly waned and herculean efforts will not be put forth to separate the two diastereomers.

5.4   Conclusion

In this chapter, the typical Group Halterman chiral bis-indenyl ligands and some corresponding metallocenes were visited. The synthesis of the 1,1'-bis-4-methylinden-7-yl binaphthyl ligand was attempted. However, the tetra-ketone formed from the Stetter reaction to the known 1,1'-binaphthyl dialdehyde appears to be unable to undergo the Erker reaction to any synthetically useful extent, if at all. While this is perhaps not too surprising considering the possibilities for internal aldol condensations, ether formations and such, it was nonetheless somewhat disappointing.

142

On a brighter note, the long awaited 1,2-bis-2-menthylinden-1-yl ethane was finally synthesized successfully. This fortunate occurrence corresponded to the use of more reactive, in situ-generated, catalytic systems formed from tris-(dibenzylideneacetone)dipalladium with either tri-o-tolylphosphino- or 2-dicyclohexylphosphino-)biphenyl. These systems, each generating significant amounts of product, combined with new work up procedures to exclude the excess Grignard materials, allowed the characterization of the ligand and its eventual metalation with zirconium and titanium.

Unfortunately, the long awaited ligand forms metallocenes that are extremely soluble. Both have resisted all attempts at crystallization to date. In fact, when 0.74 grams of the bis-2-menthylinden-1-yl zirconocene was dissolved in 0.75 milliliters of dry hexane, only faint traces of precipitate formed after three days in a low temperature freezer. Since the hoped for selectivity was not achieved, no more effort will be put into these metallocenes.

In conclusion, then, the typical Group Halterman-type chiral catalysts attempted include a binaphthyl bridged compound that was not worthwhile and an ethylene-bridged bis-2-menthylindene that was successfully metalated with both zirconium and titanium. These metallocenes are quite soluble, even in hexane. This increased solubility is probably due to the bulk of the menthyl group surrounding the zirconium center rather like tetrabutylammonium complexes with alkoxides to make them soluble in organic solvents by obscuring the charge from the surrounding aqueous solutions in phase-catalyst-modified Williamson ether synthesis [156].

143

Summary

In brief summary, the goal of the research has been primarily to generate new bis-indenyl ligands which may be metalated to generate single conformational isomers of specific intermediate conformation and to generate chiral bis-indenyl ligands of the Group Halterman family. A secondary aspect of the research has been to pioneer new routes to ligands that may be adapted to other similar moieties. Each of these goals has been achieved with varying degrees of success.

In the attempt to generate bis-indenyl ligands which can be metalated to single conformational isomers of an intermediate nature, a series of three unique bis-indenyl zirconocenes having tethers from the 7-position to the 1'- or 2'- position with ethylene and/or propylene bridges was attempted. Of the three attempted, the 7,1'-propylene-bridged zirconium dichloride was too conformationally mobile to isolate single conformational isomers of the specific diastereomers, both of which (rac- and meso-like) were formed. The 7,1'-ethylene-bridged zirconium dichloride was not achieved, but a route which could eventually lead to its formation was pioneered. The 7,2'-ethylene-bridged zirconium dichloride was successfully formed, but resisted all attempts at crystallization. However, it should be submitted for polymerization experiments and its reactivity studied in comparison to the lowest energy conformation.

In the attempt to generate chiral bis-indenyl ligands of the Group Halterman family and their corresponding metallocenes, a single successful ligand was formed with two corresponding metallocenes, the zirconocene and the titanocene. Both these

144

metallocenes are extremely soluble, even in hexane, and they resisted attempts at crystallization. However, they are substituted in the 2-position which is known to activate *ansa*-bis(indenyl)metallocenes and, therefore, should be submitted for polymerization experiments.

In addition to the successful metallocene formations that resulted from the palladium coupling, the fact that the coupling worked with the extremely hindered menthyl Grignard suggests that nearly any other Grignard can be coupled to the bis-(2-bromoinden-1-yl) ethane by these procedures. This suggestion illustrates the aspect of pioneering new routes. In fact, other members of Group Halterman are even now examining this route towards other ligands.

However, in the pioneering of new routes to bis-indenyl ligands, the work on the combined Stetter/Erker reaction may ultimately be the most valuable. With the success in this route, almost any aldehyde or α-keto-acid becomes a possible precursor for a 7-substituted indene. If another functional group in the system permits, then bis-indenyl systems are also possible, though as demonstrated by the binaphthyl attempt, not on dialdehydes to bis(indenyls) directly. This brings up a great deal of possibilities, including the use of naturally occurring aldehydes to form large, possibly chiral, substituent groups in the 7-position.

Another aspect of the possibilities from the work seen here is that of other anionic moieties, such as cyclopentadienyl, fluorenyl, indenyl and substituted versions thereof, could be used to displace the mesylate from **63** for the corresponding ligands.

In conclusion, three zirconocenes and one titanocene have been successfully

145

formed. However, the greatest benefit from the work done here is in the expanded

methodology and the plethora of new catalysts to which these routes will allow access.

146

Experimental

Except where otherwise noted, all reagents were used as purchased from commercial sources. Most common solvents (diethyl ether, THF, hexanes, toluene, benzene) were distilled under nitrogen from sodium and benzophenone in solvent stills. Dioxane was dried by storage over potassium hydroxide pellets (24+ hours) and filtered through basic alumina immediately prior to use under an inverted funnel of argon. Absolute methanol and absolute ethanol were filtered through 4 A° molecular sieves, then fractionally distilled, under nitrogen, from potassium hydroxide pellets and collected onto 4 A° molecular sieves. Triethyl amine was fractionally distilled from potassium hydroxide pellets, under nitrogen, and collected onto 4 A° molecular sieves. Methyl vinyl ketone, when purchased as a 99+% pure compound, was fractionally distilled from traces of potassium hydroxide and 4 A° molecular sieves, under nitrogen or argon, to 4 A° molecular sieves. Methyl vinyl ketone, when purchased as a 90% pure compound was mixed with anhydrous calcium chloride and anhydrous potassium carbonate (equal parts by weight) in a flask surrounded by an ice bath, allowed to warm towards room temperature, then fractionally distilled, under argon, onto molecular sieves.

All air and moisture sensitive reactions were carried out under an inert atmosphere of nitrogen or argon using standard Schlenck techniques and/or in a Vacuum Atmospheres Dri-Box (under pre-purified nitrogen). Routine solvent removal was performed on a Buchi RE-111 rotary evaporator using a house vacuum system followed by high vacuum removal of remaining solvents with an Edwards 5 high vacuum system

147

attached to a Schlenck line with liquid nitrogen trap and pre-purified argon.

All Thin Layer Chromatography (TLC) was performed on EM Science pre-prepared TLC plastic sheets (0.2 mm Kieselgel Silica gel 60 F_{254}). Rf's were measured to the nearest millimeter on 3-5 centimeter strips, so standard deviation is approximately +/- 0.06. All Silica plugs and column chromatography were performed with flash silica gel (E. Merck Reagents silica gel 60-200 mesh).

All NMR spectra were obtained with Varian XL-300 or 400 instruments with chemical shifts assigned from residual deuterated solvent peak, $CDCl_3$ unless otherwise noted. Low resolution mass spectra (EI and CI, 70 eV) were obtained using Thermo-Finnigan's Polaris Model. High resolution mass spectra (ESI) were obtained by Micromass Q-ToF.

Chromotography columns were prepared from flash silica gel (E. Merck Reagents silica gel, 230-400 mesh). Melting points were determined in Pyrex capillary tubes on a Mel-Temp apparatus and are reported uncorrected.

All ^{1}H NMR and ^{13}C NMR were obtained on 300 MHz or 400 MHz instruments. Only characteristic and/or strong signals are reported. Chemical shifts in parts per million (ppm) are reported relative to residual solvent peaks in the deuterated solvents. In ^{1}H NMR residual peak (or center of peaks) for chloroform, benzene, methylene chloride, and toluene are set at 7.24 ppm, 7.15 ppm, 5.32 ppm, 2.08 ppm respectively. For ^{13}C NMR, the residual chemical shifts for chloroform, benzene, and methylene chloride are set to 77.0 ppm, 128.0 ppm, and 55.0 ppm respectively. The interpretation of the spectra is reported as follows: chemical shift (number of hydrogen deduced from electronic

148

integration, multiplicity, coupling constants in hertz rounded to the nearest 0.5 when applicable, assignment of peak to hydrogens labeled in figure when applicable).

In general, the highest consistent yield of each reaction will be shown. The RF-# corresponds to the research notebook experiment and saved computer files. Other experiments of the same basic ilk (differing only in minor aspects) are listed in parentheses with the corresponding RF-#. The number assigned in the thesis, if applicable, is listed in bold beneath the structure.

4, 7-dioxooctanoic acid (via Stetter Reaction[98])-RF121 (117, 131, 134, 137)

2-Ketoglutaric acid (130 mmol, 18.98 g, 1 equivalent) was placed with a stir bar, in a one liter Schlenk flask with an attached water condenser and Suba-seal

1. Et_3N/Dioxane
2. Thiazolium Catalyst
3. MVK

Stetter Reaction: Formation of 1,4-diketones from oxoacids

septum. A thiazolium salt [3-benzyl-5-(2-hydroxyethyl)-4-methyl thiazolium salt (10.0 mmol, 2.69 g, 0.0800 equivalent)] was added to a sidearm addition flask and placed in the remaining opening of the Schlenk flask. The apparatus was evacuated and filled with argon multiple times. Dioxane (130.0 mL for a 1M solution in acid) was syringed into the flask with stirring, dissolving the solids to form a clear, colorless solution. The solution was stirred until all solids dissolved (~10 minutes) and triethylamine (320 mmol, 44.6 mL, 2.20 equivalents) added. A white precipitate formed immediately but was dissolved within twenty minutes of stirring. The sidearm addition flask was swivelled to add the thiazolium salt After twenty minutes of stirring, methyl vinyl ketone (MVK [10.0 mmol, 2.70 g]) was syringed into the mixture. A bubbler was attached through the septum via a lure-lock needle. The argon was turned off and the solution heated to 80 ^{0}C. A gas (presumed to be carbon dioxide) was evolved and the solution began to turn pale yellow. By morning the solution was dark brown. The solvent was removed and the

150

residue triturated with 100.0 mL petroleum ether to remove any remaining methyl vinyl

ketone (or its polymers). The remaining oily black residue was dissolved in chloroform

and washed with 20% sulfuric acid (4x), water (3x), and saturated sodium chloride (1x).

After drying with anhydrous sodium sulfate, the chloroform was removed to give the

crude product. Purification was accomplished by dissolving the crude product with

sodium hydroxide (65.0 mL of 1 M) and washing with diethyl ether (3x 100.0 mL).

Sulfuric acid (20% v/v) was added to pH < 2. The acidic solution was extracted with

copious amounts of dichloromethane. The combined extracts were dried with anhydrous

sodium sulfate, filtered and solvent removed to afford light brown crystalline 4,7-

dioxooctanoic acid (7.76 g, 35.7% yield) with the look and smell of brown sugar. No

movement is seen in any common TLC solvent. mp 54-56°C ^{1}H NMR (400 MHz,

$CDCl_3$): 2.76 (t, 2H, J=6 Hz), 2.68-2.73 (m, 4H), 2.61 (t, 2H, J=6 Hz), 2.16 (s, 3H)

151

4,7-dioxooctanoic acid (via Novák Modification[99])—RF152 (144, 159, 176, 243)

2-Ketoglutaric acid (34 mmol, 5.0 g) was placed, with a stir bar, in a Schlenk flask with an attached water condenser (with Suba-seal septum) and a sidearm addition flask charged with thiazolium catalyst (3-benzyl-5-(2-hydroxyethyl)-4-methyl thiazolium salt) (~3.40 mmol, 0.900 g) . The system was attached to an argon/vacuum line and evacuated for an hour. The apparatus was

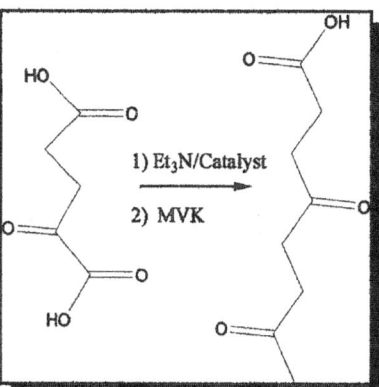

Novák Modification [99]: Stetter Reaction on Aldehydes to form 1,4-diketones using MVK as solvent

pumped and filled with argon multiple times. The flask was placed into an ice bath and allowed thirty minutes to cool. Triethylamine (~86.0 mmol, ~12.0 mL) was added via syringe. The solution was stirred 30 minutes, then methyl vinyl ketone (MVK) was syringed into mixture (~105 mmol, ~9.00 mL). The sidearm addition flask was swivelled to add in the thiazolium salt. A bubbler was attached by needle to the septum. The water bath was heated to 67 ^{0}C and stirred until carbon dioxide evolution ceased (~44 hours). The bubbler line was removed from the system and MVK removed via high vacuum. The dark brown residue was dissolved into a minimum amount of chloroform. An additional portion of chloroform (~5 mL) was added –very little insoluble precipitate remained from which to decant chloroform. One dropperful of concentrated hydrochloric was added, then 20% sulfuric acid to pH of 1. The layers were separated. The aqueous acid was extracted with an approximately equal volume of chloroform. The combined organic

152

portion was washed with ~50 mL 20% sulfuric acid, ~50 mL distilled water, and dried

over anhydrous sodium sulfate. The solvent was removed to give the acid as a light

brown solid (3.15 g, 58.9% yield). mp 55-56 °C ^{1}H NMR as above.

3-(4-methylinden-7-yl)-propanoic acid, sodium salt and isomer 3-(7-methylinden-4-yl) propanoic acid, sodium salt—RF123 (122)

A sample of 4,7-dioxooctanoic acid (7.76 g, 45 mmol) and sodium methoxide

Erker Cyclization [100]: Cyclization of 1,4-diketones to substituted indenes

(5.36 g, 99 mmol) were placed in a Schlenk flask with a stir bar. The flask was evacuated

and filled with argon. Absolute methanol was syringed into flask at 0 °C. The solution

was stirred for thirty minutes at 0 °C. Freshly cracked cyclopentadiene (~5.5 mL, ~67.5

mmol) was added by syringe and the solution was allowed to warm to and was stirred at

room temperature for at least twenty-four hours (longer neither helped nor hurt yields). A

black precipitate formed which was isolated by decantation. The remaining product was

filtered from remaining insolubles, extracted from the aqueous with diethyl ether, dried

153

(magnesium sulfate), and concentrated. The crude product was dissolved into methylene chloride and dried (sodium sulfate). Solvent was removed to afford the isomeric product (2.0 g, 22% yield). The ratio of double-bond isomers ranges from 45/55 to 60/40. Acidification of the aqueous layer with 3M HCl to ph < 2 and extraction with diethyl ether (3x, 200 mL each) afforded more of the acid. The black precipitate also afforded more of the crude acid]. ^{1}H NMR (400 MHZ, CDCl$_3$): δ 6.94-7.05 (m, 3H, H3, 5, 6); 6.55/6.58 (m, 1H, H2); 3.28/3.35 (t, 2H, J=2, H1); 3.00/3.08 (t, 2H, J=5, H10 r 12); 2.69 (m, 2H, H12 or 10); 2.33/2.41 (s, 3H, H11)

3-(4-methylinden-7-yl)-propanoic acid and isomer 3-(7-methylinden-4-yl) propanoic acid—RF139 (131, 150, 153, 155, 184, 245)

The 4,7-dioxooctanoic acid (4.26 g, 29.0 mmol) and sodium methoxide (3.44 g, 64.0 mmol) were placed in a Schlenk flask with a stir bar which was evacuated and filled with argon. After submersion in an ice bath, distilled absolute methanol (~30 mL) was syringed slowly into the flask. After stirring the solution for thirty minutes at 0 °C,

freshly cracked cyclopentadiene (~3.6 mL, ~45 mmol) was added. The solution was

154

allowed to reach room temperature and was stirred for 72 hours. The solution turned dark

brown. After acidification with 3M HCl to pH ~1, the aqueous solution was extracted

with 50 mL portions of diethyl ether until the ether extract was clear. The ether

extractions were combined, dried (sodium sulfate), and removed from the crude product

by rotoevaporation. The crude product was dissolved in a minimum amount of

chloroform and filtered. After drying with anhydrous sodium sulfate, the solvent was

removed to afford relatively pure product (3.25 g, 55% yield). $Rf_{(MeCl2)}$ = 0.05, $Rf_{(Et2O)}$ =

0.86. ^{1}H NMR (400 mHz, CDCl$_3$): 11.86 (br s, ~1H,) 6.94-7.05 (3H, m); 6.55/6.58

(1H, m); 3.28/3.35 (2 H, t, J=2); 3.00/3.08 (2 H, t, J=5); 2.69 (2H, m); 2.33/2.41 (3H,

s)

3-(7-methylinden-4-yl)propan-1-ol (Isomer A) / 3-(4-methylinden-7-yl])propan-1-ol (Isomer B) Method 1—RF124 (250, 253)

The 3-(4-[7-methyl

indenyl])-propanoic acid (and

isomer) (2.0 g, 9.9 mmol) was

placed in Schlenk flask with a

stirbar, under argon. Dry ether

(150 ml) was syringed into flask

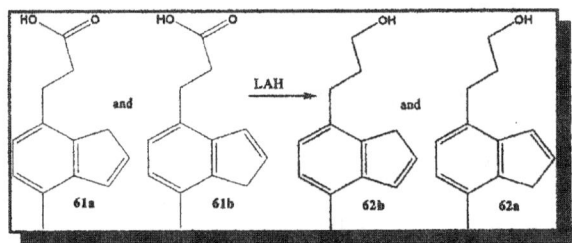

Lithium Aluminum Hydride (LAH) Reduction

and cooled to 0 °C. Lithium aluminum hydride (0.75 g, 19.9 mmol) was suspended in

155

dry ether (~50 mL) and cannulated (slowly!) into the solution. After stirring one hour

with warming towards room temperature and two hours at room temperature, the reaction

mix was quenched with distilled water (1.0 mL water per 1.0 g lithium aluminum

hydride) then sodium hydroxide (0.75

mL of 1M). Stirring was stopped to

allow precipitation of aluminum

hydroxide and water (2.25 mL) was

added to compact the solids. The ether

fraction was decanted from the

aluminum hydroxide precipitate, which

was then washed with anhydrous ether

(3 x ~50 mL each), being careful to not

break up solid. After decanting,

the ether portions were combined,

dried with anhydrous magnesium

sulfate. After solvent removal, a

clear colorless, slightly viscous

liquid resulted. (1.74 g, 93%

yield) of the clear colorless

alcohols (double-bond isomers).

$Rf_{(MeCl2)} = 0.17$, $Rf_{(Et2O)} = 0.78$. ^{1}H

NMR (300 mHz, $CDCl_3$) Isomer

Hydrogen Set	Chemical Shift	Integration	Multiplicity	Coupling Constants
A	1.2928	1H	t	$J_{AF} = 7.0$ Hz
B	1.8998	2H	tt	$J_{BD} = 10.0$ Hz, $J_{BV} = 8.5$ Hz
C	2.3348	3H	br s	
D	2.8398	2H	br t	$J_{DB} = 7.0$ Hz
E	3.2878	2H	dt (looks t)	$J_{EI} = 3.0$ Hz, $J_{EG} = 3.0$ Hz
F	3.6798	2H	dt (looks br q)	$J_{FI} = 8.5$ Hz, $J_{FA} = 7.0$ Hz
G	6.5688	1H	dt	$J_{GE} = 7.5$ Hz, $J_{GE} = 3.0$ Hz
H	6.9418	1H	br d	$J_{HI} = 10.5$ Hz
I	7.0158	1H	dt	$J_{IG} = 7.5$ Hz, $J_{IE} = 3.0$ Hz
J	7.0248	1H	br d	$J_{JH} = 10.5$ Hz

Hydrogens for Alcohol **62a**

Carbon Number	Chemical Shift	DEPT	HMQC Coupled	HMBC Coupled
1	18.6658	CH_3	C	H*
2	29.1548	CH_2	D	B, F*
3	34.3258	CH_2	B	D, F
4	38.3778	CH_2	E	G*
5	62.4348	CH_2	F	B, D
6	125.7808	CH	H	C
7	126.6258	CH	I or J	D
8	130.0628	CH	I or J	C, E
9	130.4398			C, G*
10	131.6178			B, D, H*
11	133.4028	CH	G	E
12	142.2478			C, E, H*
13	142.6638			C, D, I or J

Alcohol **62a**'s Carbon Data

156

A: 6.92-7.04(3H, m); 6.55(1H, m); 3.68(2H, m,); 3.28(2H, m); 2.83(2H, m); 2.33(3H, s); 1.84-1.97(2H, m); 1.53 (1H, s) Isomer B: 6.92-7.04(3H, m); 6.55(1H, m); 3.68(2H, m); 3.34(2H, m); 2.75(2H, m); 2.41(3H, s); 1.84-1.97(2H, m); 1.53 (1H, s) MS (EI/Dip) 188.98(24.20),187.97 (100), 154.99 (37.75), 144.00(22.39), 129.00 (21.75), 127.98 (23.49); HRMS: calculated, C13H16ONa m/e 211.1019; found, 211.1119. A sample of the alcohols with a significantly enhanced A to B ratio was obtained via silica gel column (hexane) allowing the data to the right to be obtained as well as: COSY spectrum coupling A to F, B to D and F, D to B, E to G and I or J, F to A and B, G to E and I or J, H to I or J, I/J to G and H); NOESY of (H$_C$) in **62a** on H$_H$ and H$_E$ and in **62b** on H$_H$ and H$_J$.

3-(7-methylinden-4-yl)propan-1-ol [Isomer A]/ 3-(4-methylinden-7-yl])propan-1-ol [Isomer B] Method 2–RF164

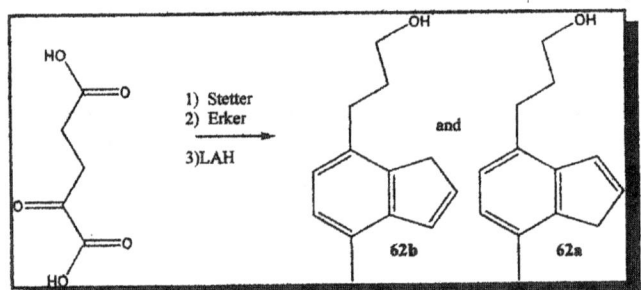

After assembling a Schlenk flask with reflux condenser, stir bar, and side arm flask, 2-ketoglutaric acid (0.978 g, 6.69 mmol) was added to the flask under argon and 3-benzyl-5-(2-hydroxyethyl)-4-methyl thiazolium salt (0.144 g, 0.540 mmol) was added to the side arm flask. High vacuum was pulled for one hour to remove any traces of moisture as the catalyst looked somewhat suspect. The flask was then filled with argon

157

and cooled in an ice bath. A bubbler line was attached to the septum of the reflux condenser via a needle and argon was bubbled through for a few seconds. The positive argon flow was stopped and freshly distilled triethylamine (2.33 mL, 16.8 mmol) was syringed into the flask with stirring. The mixture was allowed to stir in the ice bath for thirty minutes then the catalyst was added by swivelling the arm. After stirring for ten minutes, freshly distilled methyl vinyl ketone (MVK) (1.67 mL, 20.1 mmol) was added. The flask was then heated to 78°C for 15 h (overnight) with slow continuous carbon dioxide evolution. Another portion of MVK (1.0 mL) was added and heating continued at 78 °C for 8 h (carbon dioxide evolution was evidenced for about an hour after addition of extra MVK). The bubbler line was then removed and excess MVK was removed with high vacuum. The flask was then refilled with argon and placed back in the ice bath. Sodium methoxide (2.35 g, 43.6 mmol), dissolved in freshly distilled absolute methanol (~20 mL) and cooled in an ice bath, was cannulated into the reaction flask. Freshly cracked cyclopentadiene (3.00 mL) was syringed into the reaction flask . It was then allowed to warm to room temperature and stirred for 15 h (overnight). Hydrochloric acid (16.6 mL of 3M, 50.0 mmol) was used to quench the reaction. The reaction mixture was then extracted with copious amounts of diethyl ether (~100 mL) and methylene chloride (~400 mL). The organic fractions were combined, washed once with saturated sodium chloride solution (~50 mL), then dried with anhydrous sodium sulfate for 3 h in a sealed Erlenmeyer flask with stirring. The solvent was then removed by rotary evaporation and the residue subjected to high vacuum. After several hours of high vacuum, the flask was filled with argon and lithium aluminum hydride (0.75 g, ~20 mmol) was added. Dry

158

diethyl ether (~20 mL) was syringed slowly into the flask and the solution allowed to stir

12 h (overnight). Distilled water (0.75 mL) was used to quench the reaction. Sodium

hydroxide (0.75 mL of 1M) was used to form aluminum hydroxide, and distilled water

(2.25 mL) was used to compact the solids. The ether solution was then carefully decanted

and the remaining salts washed with dry diethyl ether (3x25 mL), swirling carefully to

not break up the precipitates. The combined ether solution was dried two hours over

anhydrous sodium sulfate in a sealed Erlenmeyer flask, filtered through a coarse fritted

funnel, and concentrated to a slightly yellowish oil (1.1565 g, 6.14 mmol) to afford the

alcohol in 92% yield. After a short silica gel plug, collected in three fractions: pure

alcohol was achieved as a clear colorless oil in 86 % yield (1.0841 g); mostly pure

alcohol in 4% yield (0.0644 g); and some slight impurity (0.008 g) was obtained.

Characterization was as above.

Mesylate–RF-166 (125A)

Alcohol **62**
(1.08 g, 5.76 mmol)
was placed in a
Schlenk flask which
was then vacuumed
and filled with argon.

Mesylation with Mesylchloride (MsCl)

Methylene chloride (20 mL) was syringed into the flask to dissolve the alcohol. The

159

solution was cooled to –30 °C with stirring, and freshly distilled triethylamine (0.90 mL, 6.9 mmol) was syringed into the solution and allowed to stir for 10 minutes. Methanesulfonyl chloride (MsCl) (~0.5 mL) was added neat, and the reaction mixture got quite smoky. After ten minutes at –30 °C, the orangish-brown solution was allowed to warm towards room temperature for thirty minutes, then washed with 1 M HCl, distilled water, and saturated sodium chloride solution (10 mL each). The solution was then dried with anhydrous sodium sulfate, and the solvent removed to 1.4043 g (92% yield) of dark green (almost black) oil. $Rf_{(MeCl2)} = 0.51$, $Rf_{(Et2O)} = 0.92$. . ^{1}H NMR (400 mHz, CDCl$_3$) Isomer A: 6.90-7.05 (3H, m); 6.59 (1H, dt, J = 6, 2), 4.21 (2H, t, J = 6); 3.29 (2H, t, J =2); 2.97 (3H, s); 2.87 (2H, t, J = 8); 2.33 (3H, s); 2.04-2.14 (2H, m) Isomer B: 6.90-7.05 (3H, m); 6.55 (1H, dt, J = 5.5, 2), 4.24 (2H, t, J =6); 3.32 (2H, t, J =2); 2.97 (3H, s); 2.79 (2H, t, J =7); 2.41 (3H, s); 2.04-2.14 (2H, m) ^{13}C NMR (300 mHz, CDCl$_3$): 18.749, 18.917, 28.895, 29.013, 29.757, 30.967, 38.316, 38.678, 69.662, 69.807, 124.956, 126.284, 127.085, 128.142, 128.707, 130.160, 130.405, 130.744, 131.272, 133.495, 134.220, 142.817, 143.077 (Both isomers). DEPT: CH$_3$ (18.749, 18.917), CH$_2$ (28.895, 29.013, 29.757, 30.967, 38.316, 38.678, 69.662, 69.807), CH (124.956, 126.284, 127.085, 128.142, 130.160, 130.744, 133.495, 134.220) MS: (EI/DIP 70eV)155.19(91.26) 170.19(60.12) [M]266.07(100.00) [M+1]267.07(25.17) [M+2]268.07(8.51) [M+3] 269.11(2.39) HRMS calculated mass m/e C14H19O3S 267.1055; found 267.1036.

160

Ligand 64: 1-(7-methylinden-4-yl)-3-(inden-1-yl)propane RF125B (168, 262)

Indene (0.898 g, 7.73 mmol) was dissolved in dry diethyl ether (10 mL) in a Schlenk flask with a stir bar and cooled to -60 °C. A solution of n-butyl lithium

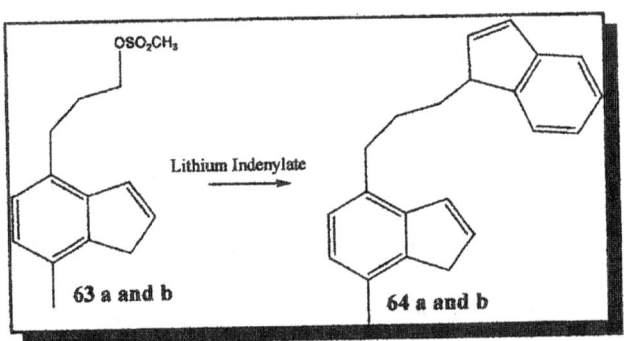

63 a and b   64 a and b

(2.60 mL, 2.6M, 6.76 mmol) was added dropwise. The resulting solution was allowed to

warm towards room temperature for fifteen minutes. The solution was then cooled to -78

°C. A sample of the mesylate (1.29 g, 4.83 mmol) dissolved in diethyl ether (~10 mL)

was syringed into the solution and the solution left overnight in the dry ice/acetone bath

to afford a very slow approach to room temperature. The resulting roseate solution was

washed twice with 1 N hydrochloric acid, water, and saturated sodium chloride solution

(10 mL each). The orange solution is dried with anhydrous sodium sulfate for three

hours, then concentrated and put through a silica gel plug (petroleum ether to remove

excess indene followed by methylene chloride to remove ligand. The solvent was

removed to afford the ligand as a clear, pale yellow paste (0.9428 g, 3.2 mmol) in 68%

crude yield. After a silica gel column purification with petroleum ether, the yield was

reduced to 43%. ^{1}H NMR (400 mHz, CDCl$_3$): 1.55-1.65(1H, m), 1.65-1.80(2H, m), 1.90-

161

2.00(1H, m), 2.32 and 2.38(3H, s and s, [double bond isomer in 2:1 ratio, respectively]),

2.60-2.70(1H, m), 2.70-2.80.25(H, m), 3.28 (2H, very fine dd, J = 1.0 HZ, J = 1.0 Hz,),

3.49(1H, dt, J = 6Hz, J = 4Hz), 6.50-6.527(1H, dd, J = 3 Hz, J = 2Hz), 6.52-6.56 (1H, dt,

J = 6 Hz, J = 1Hz), 6.75-6.80(1H, dt, J = 6Hz, J = 2Hz), 6.85-7.03(3H, m), 7.12-7.18(1H,

dd, J = 7.5 Hz, J = 7.5 Hz), 7.19-7.24(1H, dd, J = 7.5 Hz, J = 7.5 Hz), 7.30-7.34(1H, d, J

= 7.5 Hz), 7.36-7.39(1H, d, J = 7.5 HZ).

Propylene-bridged-1-(7-methylinden-4-yl)-3-(inden-1-yl)Zirconium Dichloride RF146 (126, 147, 170)

Ligand 64

(0.353 g, 0.297

mmol) in hexane

(5.0 mL) was

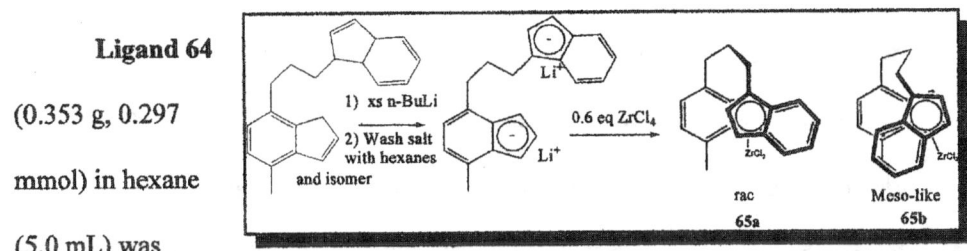

placed in a Schlenk flask with a stir bar under argon. Dry toluene (5 mL) was cannulated

into the flask under argon. After cooling to -78 °C, n-butyl lithium (0.39 mL, ~2 M, 0.78

mmol) was added. The solution was allowed to stir overnight to complete the

deprotonation. The solvent was removed from the orange solution under high vacuum

and zirconium tetrachloride was added to the flask (0.659 g, 0.283 mmol) in the

glovebox. The solids were then cooled to -78 °C by submerging the flask containing

them in a dry ice/acetone bath for thirty minutes. Solvent (~10 mL of 5:1 toluene/ether)

was added slowly to the flask and the resulting solution allowed to warm to room

162

temperature overnight with stirring. An aliquot (1.0 mL) was removed to an argon-filled

NMR tube with a vacuum adaptor and vacuumed, filled with argon, and "dissolved" in

deuterobenzene –only partially soluble. The crude ^{1}H NMR spectrum resembled the

ligand with only a few of the expected peaks of the metallocene. Repeating the ^{1}H NMR

analysis with deutero-chloroform as solvent resulted in a spectrum with significantly

more of the metallocene than ligand. The pot, therefore, was dissolved in hot methylene

chloride and filtered from the remaining zirconium salts by cannula to a new Schlenk

flask. White crystals began to settle out and the solvent was removed by vacuum to leave

the white crystals and a yellow solid. Attempted hexane washes to remove the ligand also

transferred the zirconocene. Hhot hexane (5 mL) was used to dissolve the compounds.

This mixture was evaporated to the first sign of crystals, then placed in the refrigerator.

Unfortunately, the ligand was still inseparable from the ligand. Finally, on the advice of

Vladimir Marin, a minimum amount of methylene chloride (2 mL) was used to dissolve

the most methylene chloride soluble material and the less polar ligand was allowed to

settle out of the polar solution in the freezer. This methylene chloride solution was then

transferred and concentrated to crude **RF126** (0.070 g, 82 % crude yield). While

metallocene and ligand were both still present (based on ^{1}H NMR spectra), comparison of

the spectra between the partially metalated mixture and the pure ligand allowed the

assignment of peaks to the metallocene and mole and mass ratios to be calculated. Based

on these calculations, 52% of the crude mass belongs to the metallocene for a 42% actual

yield. ^{1}H NMR (400 MHz, benzene): 1.85-205 (2 H, m), 2.10/2.30 (3H, pair of singlets),

2.40-2.50 (2H, m), 2.80/2.60 (2H, t [dd] J = 8.0 Hz), 2.90/3.0 (1H, pair of br singlets),

163

3.10/3.05 (1H, pair of br singlets), 5.90-5.95 (1H, m), 6.25-6.30 (1H, m), 6.90/6.85 (1H, m), 6.95-7.18 (2H, m), 7.15-7.25 (4H, m).

Ethyl-3-(4-methylinden-7-yl)-propionate-RF128, (156, 157, 248)

Remnants of **RF122** were triturated with ether, dried and concentrated to (0.842 g, 3.62 mmol), dissolved in ethanol (50 mL) and added

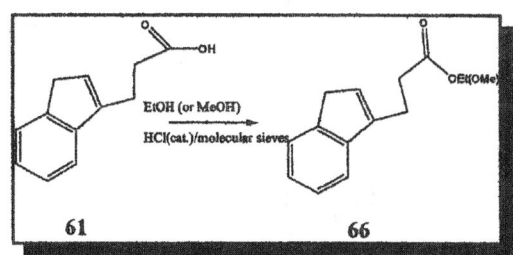

to a flask containing molecular sieves (~50 mL) under nitrogen, and attached to a Dean-Stark apparatus filled with molecular sieves and to a reflux condenser. Concentrated hydrochloric acid (2 mL) was added by syringe. The mixture was heated to reflux. By morning the reddish-brown solution and turned to a black goo. Believing the "goo" to be a bad sign, the reaction mixture was allowed to sit for nearly two weeks before finally being extracted with copious amounts of petroleum ether (~ 200 mL). The petroleum ether was then washed with a saturated sodium carbonate solution (2 x 50 mL) and water (2 x 50 mL), dried with anhydrous magnesium sulfate, and concentrated to crude product (0.775 g, 3.36 mmol). Spectral analysis was surprisingly good. A short silica plug separated some few impurities and high vacuum removed final traces of solvent, giving pure product [as a mixture of double bond isomers] (0.769 g, 3.33 mol) in 91% yield. Repeated runs run for 48 h gave 30-60% yields. ^{1}H NMR (400 MHz in CDCl3): 1.25

164

(3H, t, J = 8), 2.30 (1.74 H, s) 2.40 (1.26 H, s), 2.6 (2H, m), 2.95 (0.84 H, t, J = 9.5), 3.05

(1.16 H, t, J = 9.5), 3.25 (1.16 H, br s), 3.35 (0.84 H, br s), 4.1(2H, q, J=8), 6.50(0.42H,

br d, J =6), 6.55(0.58 H, br d, J = 6), 6.90-7.05(3H, m)

Di-Grignard Reaction to 2-(7-methylinden-4-yl)-1-(inden-2-yl-2-ol)ethane-

RF135A (128, 129, 236)

Freshly crushed

magnesium (0.972 g, 40.0 mmol)

and a large stir bar were placed in

a large argon- filled Schlenk flask

with attached reflux condenser, addition funnel, and argon outlet. Repeated evacuation

and refilling with argon established an inert atmosphere. Dry tetrahydrofuran (THF) was

added (~3 mL). Vigorous stirring for one minute resulted in shiny magnesium turnings.

1,2-Dibromoethane (~ 0.1 mL) was added and the flask heated in an oil bath until

vigorous bubbles were seen in the THF. After the bubbling ceased, fifteen minutes were

allowed to elapse to allow completion of the reaction. The flask was kept near room

temperature by submersion in a water bath. The ether was then removed by cannula and

fresh ether (~10 mL) was added. Freshly distilled α, α'-dichloro-ortho-xylene (1.49 g,

10.0 mmol) in tetrahydrofuran (136 mL) was added from an addition funnel over four

hours (~1 drop/3 s) with stirring. The solution was then stirred for sixteen hours,

resulting in a nice olive green solution. The solution was cooled to -78 °C and stirred for

165

one hour before a light tan solution of **66** (0.769 g, 3.33 mmol) in tetrahydrofuran (~20 mL) was added over two hours (1drop/2 s). When the addition was complete, the solution was allowed to warm very slowly (~2 h). The brown solution was then allowed to react at room temperature for sixteen hours with increased darkening of the brownish tint. Distilled water (~15 mL) was added to quench the reaction. On the first drop of water, the brown solution turned rather green and white precipitates began to form. By the addition of the fist milliliter the greenish solution is thick with cloudy white precipitates. By the end of the water addition the precipitates had dissolved. The stirring plate was turned off and the mixture allowed fifteen minutes to settle, but no significant layers formed, so the green solution was filtered and concentrated to a wet orange oil. This oil was dissolved in methylene chloride (~20 mL), and washed with ammonium chloride (~40 mL of 1 M). The light yellow methylene chloride solution was separated from the clear aqueous layer, washed with water (3 x ~ 40 mL), and dried with anhydrous sodium sulfate. The sodium sulfate was filtered from the methylene chloride solution, which was concentrated to crude alcohol **67**. The ^{1}H NMR spectrum showed complete consumption of starting material. ^{1}H NMR (400 MHz, CDCl$_3$): 2.0-2.2 (2H, m), 2.3-2.5 (3H, pair of singlets), 2.8-2.9 (2H, m), 2.9-3.2 (4H, m), 3.3-3.4 (2H, pair of broad singlets), 6.5-6.6 (1H, m), 6.9-7.24 (7H, m) ^{13}C NMR (300MHz CDCl$_3$) 19.5, 24.5, 39.5, 41.0, 44.0 (double intensity), 81.5, 122.0, 122.5 (double intensity), 123.0 (double intensity),124.0, 124.2 (low intensity) 125.5, 126.5, 127.0 (low intensity), 127.5(low intensity), 141.0, 141.5(low intensity) DEPT: -CH$_3$ (19.5), -CH$_2$ (24.5, 39.5, 41.0, 44.0 [double intensity]), -CH (122.0, 122.5 (double intensity), 123.0 (double

166

intensity),124.0,126.5, 141.0 (MS (Na+ technique): 185.1(58.11), 186.1(5.79), 265.1(100), 266.1(16.78), 313.2(53.44), 3134.2(12.64) HRMS: Calculated mass = 313.1568, found 313.1577.

Formation of Ligand 68: **1-(7-methylinden-4-yl)-2-(inden-2-yl)ethane Method 1-RF-135B (241)**

Crude **67** was refluxed overnight with benzene (15 mL) and p-toluenesulfonyl chloride (1.10 g) in a Dean-Stark apparatus. After cooling to

room temperature, the solution was washed with water (3x 20 mL), then saturated sodium chloride solution (2 x 20 mL). A yellowish solid precipitated from the aqueous phases and was saved. The solvent was removed to give an oily solid. The solid from the aqueous phase was combined with the oily solid and the combined solids extracted with hexane. After the removal of the hexanes and passage through a silica gel column, the isomers of pure ligand **68** remained (0.193 g, 0.710 mmol, 21% yield from the ester). ^{1}H NMR (400 MHz, CDCl$_3$): 2.3-2.5 (3H, pair of singlets), 2.8-2.9 (2H, m), 2.9-3.1 (2H, m), 3.3-3.4 (4H, pair of broad singlets), 6.6 (2H, m), 6.95-7.05(3H, m), 7.10-7.15(1H, m), 7.2-7.3(2H, m), 7.4(1H,d, J = 10) ^{13}C NMR (300 MHz, CDCl$_3$): 19.5, 32.5, 33.0, 38.5, 42.0, 120.0, 123.5, 124.0, 125.0, 126.0, 126.2, 127.8, 127.9, 130.5, 133.5, 133.8, 134.0, 142.0, 143.0, 145.0, 150.0. DEPT: CH$_3$ (19.5), CH$_2$ (32.5, 33.0, 38.5, 42.0), CH

167

(120.0, 123.5, 124.0, 125.0, 126.0, 126.2, 127.8, 127.9, 130.5, 133.5, 133.8, 134.0). MS(

70 eV): 115.2(6.94), 116.2(1.66), 143.37(100), 144.49 (23.50), 272.14 (65.18),

273.09(16.90) HRMS (Na+ Technique): Calculated 295.1463, Found: 295.1410.

Ligand 68 Formation, Method 2-RF-246 (244, 245, 247)

Alcohol **67** (0.0502 g, 0.160 mmol)
was placed in a flask with a microstir bar
and cooled to 0 °C. Triethylamine (0.40
mL), and methanesulfonyl chloride (0.20
mL) were added by syringe. The solution
was allowed to reach room temperature and

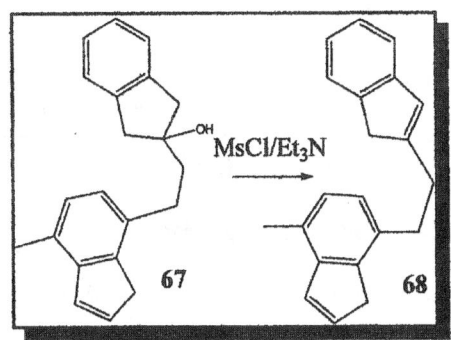

stirred overnight. After stirring overnight, the solution was heated to reflux for two

hours. Methylene chloride was added and the resulting solution washed with

hydrochloric acid (~5 mL of 1 M), saturated sodium chloride (5 mL), and water (5 mL).

After drying overnight with anhydrous sodium sulfate, the solvent was removed to a wet-

looking paste. Methylene chloride was again added and fresh magnesium sulfate

(anhydrous) was used to dry the solution, which was then filtered and subjected to high

vacuum to form a crude paste. The paste was triturated with hexane and placed on a

silica plug. The plug was flushed with hexane, removing three bands, the second of

which was crude product (0.0430 g). The plug was washed with ether and chloroform to

yield another fraction of crude product that was washed with hydrochloric acid (5 mL of 1

168

M), dried with sodium sulfate and subjected to high vacuum to yield pure product (0.0031

g) (Total: 0.0461g, 16.0 mmol, 94 % yield). Spectra as above.

Zirconocene 69: Ethylene-bridged-1-(7-methylinden-4-yl)-2-(inden-2-yl)zirconium dichloride-RF148 (249)

Ligand 68 (0.193 g, 0.700 mmol) was dissolved in dry hexane (5 mL) and cannulated to an argon-flushed flask containing a stir

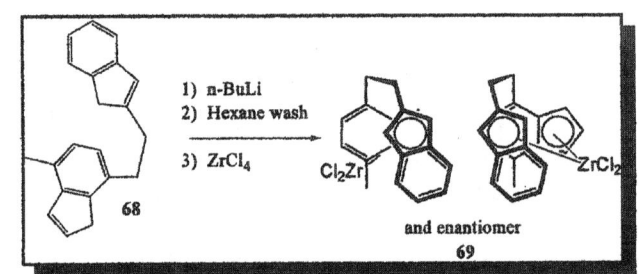

bar. The solution flask was placed in a dry ice/acetone bath at -78 °C. After stirring the

solution for twenty minutes, n-butyl lithium (~1.6 mmol, ~0.80 mL, ~2.0 M in hexane)

was added. The solution was allowed to reach room temperature and left overnight to

complete the deprotonation. The solvent was removed from the yellowish-orange

solution and the resultant salt washed with hexane by cannulae filtration. Zirconium

tetrachloride (0.132 g, 0.567 mmol) was added in the glovebox. A solvent system,

composed of 10 parts toluene to 1 part tetrahydrofuran, was added at -78 °C to dissolve

the salts (~10 mL). The solution was allowed to stir at room temperature overnight with a

resultant change of the solution color to a light orange. The reaction was allowed to stir

two more days at room temperature. The solvent was removed to afford a pasty tan solid.

By ^{1}H NMR spectroscopy it was determined that 57% of the ligand was converted to

169

metallocene. While the remaining dilithium salt was unable to be removed from the metallocene, the peaks belonging to the metallocene were able to be determined by comparison with the spectrum of the dilithium salt. ^{1}H NMR (300 MHz, CD$_2$Cl$_2$): 2.3 (3H, br s), 2..9 (2H, m), 3.0 (2H, m), 3.3-3.4 (3H, m), 6.6 (2H, m), 6.9-7.8 (6H, m).

Palladium Coupling to Form ansa-1,2-Bis-(2-menthylinden-1yl)-ethane-RF260 (203, 207, 208, 211, 214 , 239, 257, 258, 260, 261, 265, 266)

In the glovebox, tris(dibenzylideneacetone)-dipalladium (0.055 g, 0.06 mmol, [5 mol%]) and tri-o-tolyl phosphine (0.044 g, 0.144 mmol,

[12 mol %]) were placed in a Suba-sealed round bottom flask containing a stir bar. Also in the glovebox, zinc bromide (0.838 g, 3.72 mmol) was added to a Schlenk flask with a stir bar. Each flask was wrapped with foil and flushed with argon. To the Schlenk flask containing zinc bromide, tetrahydrofuran (THF) (10 mL) was added. After cooling the zinc bromide slurry to 0 °C, menthylmagnesium chloride (3.6 mL of 1M in THF), also at 0 °C, was added. The mixture was stirred until the ice had melted and the bath temperature neared room temperature (~3 h). Meanwhile, tetrahydrofuran (5 mL) was slowly added to the flask containing the catalyst mixture at 0 °C. At the end of the third hour, the catalytic solution was cannulated to the Grignard solution. The empty flask was

170

washed with tetrahydrofuran (5 mL) and the wash cannulated to the stirring solution.

After another two hours of stirring, bis-(2-bromoinden-1-yl)ethane (0.50 g, 1.20 mmol) in

tetrahydrofuran (5 mL) was added to the pot via cannula. Again, the empty flask was

washed with tetrahydrofuran (5 mL) and the wash cannulated to the reaction flask. The

mixture was allowed to react four days at room temperature, then a positive argon flow

was used to remove the solvents. Dry hexane (25 mL x2) was used to extract the crude

product from remaining Grignard salts. The golden solution was concentrated to crude

product (0.3976 g, 0.74 mmol, 51% yield). The crude product was purified via a silica

gel column with hexane as effluent to get fairly pure RF260 (0.2508 g, 0.47 mmol, 39%

yield, ~80% pure by spectral integrations). ^{1}H NMR (300 MHz, CD$_2$Cl$_2$): 0.6-1.9 (~35

H, m), 2.1-2.5 (1H, m), 2.55-2.65 (1H, m), 2.8-2.9 (4H, m), 3.1-3.4 (2H, m), 3.55-3.65

(2H, m), 7.0-7.5 (8H, m).

Ethylene-bridged-1,2-bis(2-menthylinden-1yl)zirconium dichloride-RF268,

Rf266

A sample of ligand **79**
was converted to and saved
as the dilithium salt by
treatment with n-butyl
lithium. A portion (0.342 g,
0.640 mmol) of this dilithium

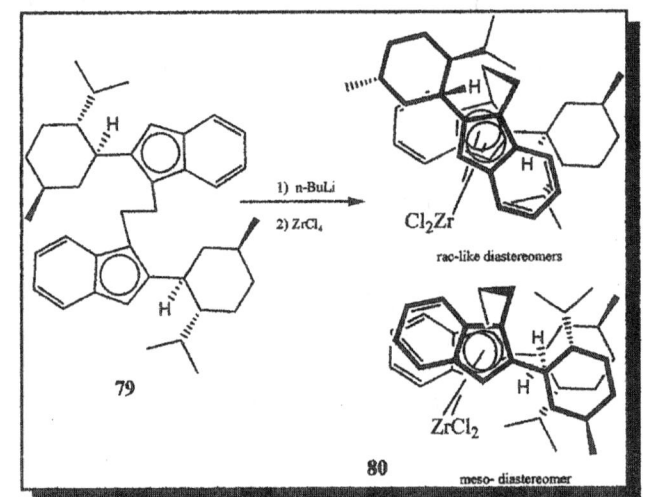

171

salt, and zirconium tetrachloride (0.149 g, 0.640 mmol) were added to a round bottom

flask in the glovebox. After flushing the flask with argon, the salts were cooled -78 °C

and toluene (~10 mL) was added dropwise. The resulting solution was stirred for 48 h at

room temperature and the toluene soluble material cannulated to three flasks. The

toluene was removed from each flask to yield crude metallocene (3 fractions totaling

1.043 g, 0.150 mmol, 23 % yield). Repeated attempts at recrystallization failed (among

the solvents attempted were hexane, toluene, benzene, and methylene chloride: in each

case, the metallocene was dissolved in a minimum amount of solvent and placed in the

freezer). A small fraction of the metallocene (0.74 g) was dissolved in a minimum

amount of hexane (0.75 mL), but only wispy hints of crystalline material resulted after

several days in the freezer. Based on the expansion of the peak at δ5.35 in deuterated-

methylene chloride (corresponding to the peak at 6.0 in deuterated-benzene, but clearer in

the methylene chloride), the presence of all three isomers (two rac-like isomers and the

meso-isomer) was confirmed in approximately a 2:2:1 ratio, respectively. ^1H NMR (300

MHz, C_6D_6): 0.6-1.9 (34 H, m), 1.9-2.3 (2H, m), 2.3-2.7 (2H, m), 2.7-3.0 (2H, m), 3.0-3.2

(2H, m), 6.0 (1H, m), 6.5(1H, m) 7.0-7.5 (8H, m); ^{13}C NMR (300 MHz, C_6D_6): 14.10,

16.261, 21.942. 22.831, 23.087, 23.175, 24.950, 28.441, 30.241, 30,581, 31.978,32.372,

33.299, 35.490, 35.955, 36.648, 37.919, 40.768, 43.801, 47.112, 47.175, 118.548,

119.182, 123.671, 123.982, 124.804, 129.012, 142.938, 142.938, 144.549 (missing one

hydrocarbon and one aromatic carbon, presumably quaternary ones).

172

Ethylene-bridged-1,2-bis(2-Menthylinden-1yl)titanium dichloride-RF268

RF269

The dilithium salt of **79** was formed by addition of n-butyllithium in hexane. The solvent was removed by cannula and the resulting solid evacuated and filled

with argon. In the glovebox, titanium trichloride (0.0633 g, 0.38mmol) was placed in the Schlenk flask containing the dilithium salt. After flushing with argon and cooling to -78 °C, toluene (~10 mL) was added. The resulting solution was stirred for 18 hours at room temperature and an additional 18 hours at 50 °C, then cooled to -78 °C. Chloroform (~10 mL) and hydrochloric acid (~10 mL of 6 M) were carefully added and the reaction warmed to room temperature. Compressed air was bubbled through the solution for 20 minutes. The aqueous layer was separated and extracted with methylene chloride (3 x 5 mL). The combined organic fractions were dried with anhydrous calcium chloride, filtered, and removed to afford crude metallocene. The flask containing the crude material was flushed with argon. This crude metallocene was dissolved in a minimum amount of chloroform and placed overnight in the freezer. Precipitates were filtered and

173

the chloroform removed to give fairly pure titanocene (0.197 g, 0.320 mmol, 84% yield [not counting the 0.0301 g of crude material used in ^{1}H NMR]). Again, as experienced with the zirconocene made from this ligand, excessive solubility prevented recrystallization. Also, the presence of all three isomers was determined by expanding the peak at δ5.35 in deuterated methylene chloride and examining the ratio of the finer peaks to find that the meso formed 26% of the mixture, one of the rac-like isomers formed 64% of the mixture and the other rac-like isomer formed the remaining 10% of the mixture. Which rac-isomer formed the majority of the mixture could not be determined. A short column seems to have caused decomposition of the titanocene. ^{1}H NMR (300 MHZ, CDCl$_3$): 0.6-1.9 (34 H, m), 2.3-2.5 (1H, m), 2.6-3.0 (2H, m), 3.1-3.4 (1H, m), 4.9-5.2 (1H, m), 6.2-6.5 (1H, m), 7.0-7.5 (8H, m).

174

References:

1. a) "Chiral Titanocens and Zirconocenes in Synthesis" Hoveyda, A. H.; Morken, J. P. in Metallocenes Togni, A.; Halterman, R. L. eds, Wiley-VCH **1998,** 625-683. (b) "Enantioselective C-C and C-H Bond Formation Mediated or Catalyzed by Chiral ebthi Complexes of Titanium and Zirconium" Hoveyda, A. H.; Morken, J. P. *Angew. Chem., Int. Ed. Engl.* **1996**, 35, 1262-1284. (c) "Synthesis and Reactivity of Substituted Cyclopentadienylmetals" Halterman, R. L. NSF Grant Proposal **2000,** 3.

2. (a) "Asymmetric Catalysis of the Diels-Alder Reaction Using Dicationic Zirconocene Complexes" Bondar, G. V.; Aldea, R.; Levy, C. J.; Jaquith, J. B.; Collins, S. *Organomet.* **2000,** 19(6), 947-949. (b) "Zirconocene-Catalyzed Cationic Diels-Alder Reactions" Wipf, P.; Xu, W. *Tetrahedron* **1995**, 51(15), 4551-62. (c) "Asymmetric Induction in the Diels-Alder Reaction Using Chiral Metallocene Catalysts" Hong, Y.; Kuntz, B. A; Collins, S. *Organomet.* **1993**, 11(11), 3600-7. (d) "[(3,5-$(CF_3)2C_6H_3)B$]$^-$[$H(OEt_2)_2$]$^+$: A Convenient Reagent for Generation and Stabilization of Cationic, Highly Electrophillic Organometallic Complexes" Brookhart, M.; Grant, B.; Volpe, A. F. Jr. *Organomet.* **1992,** 11, 3920-3922.

3. (a) "Hydrogenation of Non-functionalized Carbon-Carbon Double Bonds" Halterman, R. L. eds: Jacobsen, E. N.; Pfaltz, A.; Yamamotot, H. *Compr. Asymmetric Catal. I-III* **1999**, 1, 183-195. (b) "Asymmetric Hydrogenation of Unfunctionalized Tetrasubstituted Olefins with a Cationic Zirconocene Catalyst"

175

Troutman, M. V.; Apella, D. H.; Buchwald, S. L. *J. Am. Chem. Soc.* **1999**, 121, 4916-4917. (c) "A Designed, Enantiomerically Pure, Fused Cyclopentadienylligand with C_2-Symmetry: Synthesis and Use in Enantioselective Titanocene-catalyzed Hydrogenations of Alkenes" Halterman, R. L.; Vollhardt, K. P. C.;

Welker, M. E.; Blaser, D.; Boese, R. *J. Am. Chem. Soc.* **1987**, 109,8105-8107.

4. (a) "Stereoselective Reduction of ß-Hydroxy Ketones with Aldehydes via Tishencko Reactions Catalyzed by Zirconocene Complexes" Umekawa,Y.; Sakaguchi, S.; Nishiyama, Y.; Ishii, Y. *J. Org. Chem.* **1997**,62 (10), 3409-3412.

(b) "Enantioselective Ketone Reduction Catalyzed byChiral Titanocene Complexes" Halterman, R. L.; Ramsey, T. M.; Chen,Z. *J. Org. Chem.* **1994**, 59, 2642-2644. (b) "Catalytic Asymmetric Hydrogenation of Imines with a Chiral Titanocene Catalyst: Scope and Limitations" Willoughby, C. A.; Buchwald, S. L. *J. Am. Chem. Soc.* **1994**, 116, 8952-8965. (c) "Synthesis of Substituted Pyroles via Zirconocene Complexes of Imines" Buchwald, S. L.; Wannamaker, M. W.; Watson, B. T. *J. Am. Chem. Soc.* **1989**, 111, 776-777.

5. (a) "Generation of Perfluoropolyphenylene Oligiomers via Carbon-Fluorine Bond Activation by $Cp_2Zr(C_6F_5)_2$: A Dual Mechanism Involving a Radical Chain and Release of Tetrafluorobenzyne" Edelbach, B. L.; Bradley, M. K.; Jones, W. D. *J. Am. Chem. Soc.* **1999**, 121, 10327-10331. (b) "Activation of Carbon-Fluorine Bonds by Metal Complexes" Kiplinger, J. L.; Richmond, T. G.; Osterberg, C. E. *Chem. Rev.* **1994**,

176

94, 373-431. (c) "Carbon-Fluorine Bond Cleavage by Zirconium Metal Hydride Complexes" Edelbach, B. L.; Rahman, A. K. F.; Lachicotte, R. J.; Jones, W. D. *Organomet.* **1999**, 3170-3177.

6. (a) "Reactive Intermediates of the Catalytic Carbomegnesation Reaction: Isolation and Structures of [Cp2ZrEt]2(μμ-ethene), [Cp2Zr(ethene)(L)] (L = THF, Pyridine), and [(indenyl)₂Zr(ethene)(THF)] and of Metallacycles with Norbornene" Fischer, R.; Walther, D.; Gebhardt, P.; Göörls, H. *Organomet.* **2000**, 19, 2532-2540. (b) "Some Novelties in Olefin Carbometalation assisted by alkylmagnesium and - aluminum Derivitives and Catalyzed by Zirconcium and Titanium complexes" Dzhemileb, U. M.; Voshtrikova, O. S. *J. Organomet. Chem.* **1985**, 285(1-3) 43-51. (c) "Enantio- and Diastereoselective Catalytic Carboalumination of 1-Alkenes and alpha, gamma-Dienes with Cationic Zirconocenes: Scope and Mechanism" Shaughnessy, K. H.; Waymouth, R. M. *Organometallics* **2000**, 19, 1870-1878. (d) "Zirconium-Catalyzed Enantioselective Methylalumination of Monosubstituted Alkenes" Kondakov, D. Y.; Negishi, E.I. *J. Am. Chem. Soc.* **1995**,   117(43), 10771-2.

7. (a) "Ring Construction by Zirconium-Promoted Reductive Coupling of Nitriles with Alkenes" Mori, M; Uesaka, N.; Shihbasaki, M. *J. Chem. Soc., Chem. Commun.* **1990**, 18, 1222-4. (b) "Cyclization of Diactylenes to E,E exocyclic Dienes. Complimnetary Procedures Based on Titanium and Zirconium Reagents" Nugent, W. A.; Thorn, D. L.; Harlow, R. L. *J. Am. Chem. Soc.* **1987**, 109(9), 2788-96. (c) "*ansa*-Metallocene Derivatives. Part XLII. Cyclomagnesastion of

177

dienes catalyzed by a chiral ansa-zirconocene" Martin, S.; Brintzinger, H.H. *Inorg.*
Chem. Acta **1998**, 280(1-2), 189-192. (d) "Zirconium-catalyzed Carboalumination
of Alkynes and Enynes as a Route to Aluminacycles and their Conversion to Cyclic
Organic Compounds" Neigishi, E.I.;

Montchamp, J.L.; Anastasis, L.; Elizarvo, A.; Choueiry, D. *Tetrahedron Lett.*
1998, 39(17), 2503-2506.

8.   (a) "Some Novelties in Olefin Carbometalation Assisted by Alkylmagnesium and -
aluminum Derivatives and Catalyzed by Zirconium and Titanium Complexes"
Dzhemilev, U. M.; Vostrikova, O. S. *J. Organomet. Chem.* **1985**, 285(1-3), 43-51.
(b) "Hydromagnesation of Unsaturated Compounds by Diethylaminomagnesium
Hydride Catalyzed by Transition Metal Complexes" Dzhemilev, U. M.;
Vostrikova, O. S.;

Sultanov, R. M. *Izv. Akad. Nauk SSSR, Ser. Khim.* **1983**, 8, 1901-3. (c) "New
Reactions of alpha-Olefins with Diethylmagnesium Catalyzed by
Biscyclopentadienyl Zirconium Chloride" Dzhemilev, U. M.; Vostrikova, O. S.;
Sultanov, R. M. Izv. *Akad. Nauk SSSR, Ser. Khim.* **1983**, 1, 218-220.

9.   (a) "Zirconium-catalyzed Ethylmagnesiation of Hydroxylated Terminal Alkenes;
A Catalytic and Diastereoselective Carbon-Carbon Bond-Forming Reaction"
Houri, A. F.; Didiuk, M. T.; Xu, Z.; Horan, N. R.; Hoveyda, A. H. *J. Am. Chem.*
Soc. **1993**, 115(15), 6614-24.

10.   "Enanto-, Diastereo-, and Regioselective Zirconium-Catalyzed Carbomagnesation
of Cyclic Ethers with Higher Alkyls of Magnesium. Utility in Synthesis and

178

Mechanisitic Implications" Didiuk, M. T.; Johannes, C. W.; Morken, J. P.; Hoveyda, A. H. *J. Am. Chem. Soc.* **1995**, 117(27), 7097-104.

11. "Enantio- and Diastereoselective Catalytic Carboalumination of 1-Alkenes and α, ῶ-Dienes with Cationic Zirconocenes: Scope and Mechanism" Shaughnessy, K. H.; Waymouth, R. M. *Organometallics* **1998**, 17, 5728-5745.

12. "Remarkably Pair- and Regioselective Carbon-Carbon Bond-Forming Reaction of Zirconacyclopentane Derivatives with Grignard Reagents" Takahashi, T.; Seki, T.; Nitto, Y.; Saburi, M.; Rousset, C. J.; Negishi, E. *J. Am. Chem. Soc.* **1991**, 113(16), 6266-8.

13. "Zirconium-catalyzed Diene and alkyl-alkene coupling reactions with Magnesium Reagents" Knight, K. S.; Waymouth, R. M. *J. Am. Chem. Soc.* **1991**, 113(16), 6268-70.

14. "The Mechanism of the Zirconium-catalyzed Carbomagnesiation Reaction. Efficient and Selective Catalytic Carbomagnesiation with Higher Alkyls of Magnesium" Hoveyda, A. H.; Morken, J. P.; Houri, A. F.; Xu, Z. *J. Am. Chem. Soc.* **1992**, 114(17), 6692-7.

15. "Reactive Intermediates of the Catalytic Carbomagnesation Reaction: Isolation and Structures of [Cp2ZrEt]2(μμ-ehtene), [Cp2Zr(ethene)(L)] (L = THF, Pyridine), and [(indenyl)2Zr(ethene)THF] and of Metallacycles with Norbornene" Fischer, R.; Walther, D.; Gebhardt, P.; Göörls, H. *Organometallics* **2000**, 19, 2532-2540.

16. (a)"Zirconium-catalyzed Carboalumination of Alkynes and Enynes as a Route to

Aluminacycles and their Conversion to Cyclic Organic Compounds" Negishi, E.-I.;

Montchamp, J.L.; Anastasis, L.; Elizarvo, A.; Choueiry, D. *Tet. Lett.* **1998**, 39(17),

2503-2506. (b) "On the Regiochemistry of Cyclialkylation of Regiodefined 4-Halo-

1-Alkenylmetals Producitng Cyclobutenes" Liu, F.; Negishi, E.I. *Tet. Lett.* **1997**,

38(7), 1149-52.

17.   "Metal-promoted Cyclization. 11. Reaction of Zirconcene Dichloride with

Alkyllithiums or Alkyl Grignard Reagents as a Convenient Method for Generating

a Zirconocene Equivalent and its use in Zirconium-promoted Cyclization of

Alkenes, Alkynes, Dienes, Enynes, and Diynes" *Tetrahedron Lett.* **1986**, 27(25),

2829-32.

18.   "Zirconocene-Mediated Cyclization of 2-Bromo (sic)alpha, gama-Dienes"

Millward, D. B.; Waymouth, R. M. *Organometallics* **1997**, 16, 1153-1158.

19.   "The First Formal Asymmetric Synthesis of Phorbol" Wender, P. A.; Rice, K. D.;

Schnute, M. E. *J. Am. Chem. Soc.* **1997**, 119, 7897-7898.

20.   (a) "Zirconium" The Columbia Electronic Encyclopedia, Sixth Edition. Columbia

University Press. http://www.encyclopedia.com/articles/14188.html. (b)

"Titanium" The Columbia Electronic Encyclopedia, Sixth Edition. Columbia University

Press. http://www.encyclopedia.com/articles/12908.html.

21.   Boor, John Jr. Ziegler-Natta Catalysts and Polymerizations.. Academic Press:

New York, **1979**. p.19-29.

22.   Billmeyer, Fred W. Jr. Textbook of Polymer Science, Third Edition. John Wiley

and Sons: New York, **1984**. p.3-4.

23.   Billmeyer, Fred W. Jr. <u>Textbook of Polymer Science, Third Edition</u>. John Wiley and Sons: New York, **1984**. p.10.

24.   (a) Boor, John Jr. <u>Ziegler-Natta Catalysts and Polymerizations</u>. Academic Press: New York, **1979**. p.21. (b) The first report of lithium aluminum hydride: "Lithium Aluminum Hydride, and Lithium Gallium Hydride, and Some of their Applications in Organic and Inorganic

Chemistry" Finholt, A. E.; Bond, A. C., Jr.; Schlesinger, H. I. *Chem. Abstracts: J. Am. Chem. Soc.* **1947**, 69, 1199-1203.

25.   Bhaduri, Sumit and Doble, Mukesh. <u>Homogeneous Catalysis: Mechanisms and Industrial Applications</u>. Wiley-Interscience: New York, **2000**. p.105.

26.   "Soluble Friedel-Crafts Catalysts" Kraus, C. A.; Calfee, J. D. (Standard Oil Development Co.) *Chem. Abstract.* US 2440750 19480504 US CAN 42:34377 AN **1948**:34377.

27.   "High Molecular Weight Polymers from Propylene and 1-Butene" Fontana, C. M.; Herold, R. J.; Kinney, E. J.; Miller, R. C. (Socony-Vacuum Labs) *Chem Abstracts: Ind. Eng. Chem.* **1952**, 44, 2955-62.

28.   "Vinyl Alkyl Ethers" Schildknecht, C. E.; Zoss, A. O.; McKinley, C. (General Aniline and Film Corp.) *Chem Abstracts: Ind. Eng. Chem.* **1947**, 39, 180-6.

29.   "Polyvinyl Isobutyl Ethers-Properties and Structures" Schildknect, C. E.; Gross, S. T.; Davidson, H. R.; Lambert, J. M.; Zoss, A. O. (General Aniline and Film Corp.)

181

Chem Abstracts: Ind. Eng. Chem. **1948**, 40, 2104-15.

30.   "Crystalline High Polymers of alpha-Olefins" Natta, G.; Pino, P.; Corradini, P.;

Danusso, F.; Mantica, E.; Mazzanti, G.; Moraglio, G. Montecatini, M. *J. Am.*

Chem. Soc. **1955**, 77, 1708-10.

31.   "Ionic Polymerizations of Some Vinyl Compounds" Schildknecht, C. E.; Zoss, A.

O.; Grosser, F. (General Aniline and Film Corp.) *Chem Abstracts: Ind. Eng. Chem.*

1949, 41, 2891-6.

32.   Billmeyer, Fred W. Jr. <u>Textbook of Polymer Science, Third Edition</u>. John Wiley

and Sons: New York, **1984**. p.91.

33.   Note: His publication list dates back to at least the **1920**'s.

34.   "Twenty-five Years of 'Contributions to the Knowledge of Trivalent Carbon'"

Ziegler, Karl. Angew. *Chem.* **1949**, 61, 168-79.

35.   "Formation Mechanism and Structure of Butadiene Polymers" Ziegler, K. *Chem.*

Abstracts: Kunstoffe **1949**, 39, 45-6.

36.   "New Types of Organic Catalysts" Ziegler, K. *Chem. Abstracts: Brennstoff-Chem.*

1949, 30, 181-4.

37.   "Stereoisomerism of Eight-membered Cyclo.ovrddot.olefin Rings" Ziegler, K.

Wilms, H. *Chem. Abstracts: Naturwissenschaftten* **1948**, 35, 157-8.

38.   "Trivalent Carbon. XXII. Tetraalkyldiphenylethanes" Ziegler, K.; Deparade, W.

Chem. Abstracts: Ann. **1950**, 567, 123-41.

182

39. "Trivalent Carbon. XXIII. The Decompostion of Azoisobutyronirile and Related Substances" Ziegler, K.; Deparade, W.; Meye, W. *Chem. Abstracts: Ann.* **1950**, 567, 141-51.

40. "Trivalent Carbon. XXIV. Excitation of Polymerization by Radicals" Ziegler, K.; Deparade, W.; Kulhorn, H. *Chem. Abstracts: Ann.* **1950**, 567-151-79.

41. "Polymembered Ring Systems. XIII. Eight-membered cyclo.ovrddot.olefin rings" Ziegler, K.; Wilms, H. *Chem. Abstract: Ann.* **1950**, 567, 1-43.

42. "Butadiene and its Polymerization" Ziegler, K.; Eimers, E.; Hechelhammer, W.; Wilms, H. *Chem. Abstract: Ann.* **1950**, 567, 43-96.

43. "Organoalkali Compounds. XVI. The Thermal Stability of Lithium Alkys" Ziegler, K.; Gellert, H. G. *Chem. Abstract: Ann.* **1950**, 567, 179-85.

44. "Organoalkali Compounds. XVII. Reactions between Lithium Alkyls and Ethers" Ziegler, K.; Gellert, H. G. *Chem. Abstract: Ann.* **1950**, 567, 185-95.

45. "Organoalkali Compounds. XVIII. Addition of Lithium Alkyls to Ethylene" Ziegler, K.; Gellert, H. G. *Chem. Abstract: Ann.* **1950**, 567, 195-203.

46. "Alumino-Organic Syntheses in the Field of Olefinic Hydrocarbons" Ziegler, K. Angew. *Chem.* **1952**, 64, 323-9.

47. "Organoalkali Compounds. XIX. Reactions of the Aluminum-hydrogen Linkage with Olefins" Ziegler, K.; Gellert, G. H.; Martin, H.; Nagel, K.; Schneider, *J. Chem Abstracts: Ann.* **1954**, 589, 91-121.

183

48. Note: Source is Scifinder Scholar author search, refined by patents. The patents from which information was gleamed via their abstracts include, in no particular order: (a) "Trialkylaluminum Compounds" Ziegler, K.; Zosel,K. US 2691668 **1954** 1012. (b) "Polymerization of Ethylene" Ziegler, K. GB 713081 19540804. (c) "Polymerization of Ethylene" Ziegler, K.; Gellert, G. US 2699457 **1955** 0111. (d) "Trialkylaluminum Compounds" Ziegler, K. GB 772174 19570410. (e) "Polymerization of Ethylene in the Presence of an Aluminum Trialkyl Catalyst" Ziegler, K. GB 777152 **1957** 0619. (f) "High Molecular-Weight Polyethylenes and Catalysts for their Preparation" Ziegler, K. GB 829627 **1960** 0302. GB CAN 54:84001. (g) "Polymerization of Ethylene in the Presence of an Aluminum Trialkyl Catalyst" Ziegler, K.; Wilke, G.; Holzkamp, E. US 2781410 **1957** 0212 US CAN 51:49814. (h) "Polymerization of Olefins to Polymers of High Molecular Weight" Ziegler, K. GB 810023 **1959** 0311 GB CAN 53:128619.

49. Boor, John Jr. Ziegler-Natta Catalysts and Polymerizations. Academic Press: New York, **1979**. p22-4.

50. "Polymerization of Ethylene and Other Olefins" Ziegler, K.; Holzkamp, E.; Breil, H. *Chimica e Industria* **1955**, 37, 881-2.

51. "Organoalkali Compounds. XXI. Metal Compounds of Cyclopentadiene" Ziegler, K.; Froitzheim-Kuhlhorn, H.; Hafner, K. *Chem. Abstracts*: Chem. Ber. **1956**, 89, 434-43.

52. "New Developments in Organometallic Chemistry" Ziegler, K. *Angew. Chem.*

1956, 68, 721-9.

53.　"Crystalline High Polymers of alpha-Olefins" Natta, G.; Pino, P.; Corradini, P.; Danusso, F.; Mantica, E.; Mazzanti, G.; Moragli, G.; Montecatini, M. *J. Am. Chem. Soc.* **1955**, 77, 1708-10.

54.　(a) "Isotactical Polymers" Natta, G.; Politecnico, M. *Chem. Abstract; Chimica e Industria* **1955**, 37, 888-900. (b) "Syndiotactic Polypropylene" Nata, G.; Zambelli, A.; Pasquon, I. *Chem . Abstract of Patent* **1967**, USXXAM US 3335121 19670808. US 19630703. CAN 67:100577.

55.　"The Nobel Prize in Chemistry 1963" Professor A. Fredga. http://www.nobel.se/chemistry/laureates/1963/index.html

56.　Billmeyer, Fred W. Jr. Textbook of Polymer Science, Third Edition. John Wiley and Sons: New York, **1984**. p.91-2.

57.　Bhaduri, Sumit and Doble, Mukesh. Homogeneous Catalysis: Mechanisms and Industrial Applications. Wiley-Interscience: New York, **2000**. p33.

58.　"Synthesis of Chiral Titanocene and Zirconocene Dichlorides" Halterman, R. L. In Metallocenes: Synthesis, Reactivity, Applications Volume 1, Eds Antoio Togni and Ronald L. Halterman. p 455-544.

59.　"Synthesis and Applications of Chiral Cyclopentadienylmetal Complexes" Halterman, R. L. *Chem. Rev.* **1992**, 92, 965-994. Note: This article is a comprehensive treatise on cyclopentadienyls.

185

60. "Organometallic Compounds" National Lead Company of Britain. *Chem. Abstracts* GB 798001 **1958** 0709 GB CAN 53:11858.

61. "Polymerization of Olefins" Stuart, A. P. *Chem. Abstracts* US 2914515 **1959** 1124 US CAN 54:26375.

62. "A New Preparation of Dicyclopentadienyl Compounds" Birmingham, J. M.; Seyferth, D.; Wilkinson, G. J. Am. Chem. Soc. **1954**, 76, 4179. And references therein.

63. "Organozirconium and Organovanadium Compounds Used for the Polymerization of Ethylene" Breslow, D. S. *Chem Abstracts* US 2924593 **1960** 0209 US CAN 54:59737.

64. "Complex Bimetallic Cyclopentadienyl Organometallic Compounds" Shapiro, H.; De Witt, E. G.; Brown, J. E. *Chem. Abstracts* US 3030398 **1962** 0417. US 19590803.

65. Ege, Seyhan. <u>Organic Chemistry: Structure and Reactivity, fourth edition</u>. Houghton Mifflin: New York, **1999**. p.200-1.

66. Solomons, T. W. G.; Fryhle, C. B. <u>Organic Chemistry, seventh edition</u>. John Wiley and Sons, Inc: New York, **2000**. p. 193.

67. "Stereochemical Studies. III. The Problem of Optically Active Alkali Organic Compounds" Ziegler, K.; Wenz, A. *Chem. Ber.* **1950**, 83, 354-8.

68. For reviews of metal-centered chirality, Halterman [58] suggests Brunner, H. *Acc. Chem. Res.*. **1979**, 12, 250 and *Adv. Organometal. Chem.* **1980**, 18, 151.

186

69.   (a) "Crystal Structure and Relative Configuration of a Titanocene Complex Having a Planar Chirality and a Chirality Centered on the Titanium Atom" Lecomte, C.; Dusausoy, Y.; Protas, J.; Tirouflet, J.; Dormaond, A. Nancy, Fr. *J. Organomet. Chem.* **1974,** 73(1), 67-76. (b) "Anti-Inflammatory Planar Chiral [2,2]paracyclophaneacetic acid enantiomers" Imming, P.; Graf, M.; Tries, S.; Hirschelmann, R.; Krause, E.; Pawlitzki, G. *Inflammation Research.* **2001,** 50(7), 371-4.

70.   "Zirconium Complexes with Metallic Chirality" Tainturier, G.; Gautheron, B.; Renaut, P.; Etievant, P.; Gabriel, D. Fr. C. R. *Chem. Abstracts: Hebd. Seances Acad. Sci., Ser. C* **1975,** 281(23), 1035-6.

71.   "Titanocene Dichloride Complexes Having Central Metal Atom Pseudoasymmetry" Leblanc, J. C.; Moise, C.; Gabriel, D. Fr. *J. Organomet. Chem.* **1977,** 131(1), 35-42.

72.   "Chiral Cyclopentadienyl Ligands in the Asymmetric Hydrogenation in Homogeneous Phase of 2-phenyl-1-butene" Cesarotti, E.; Vitiello, R. *Chem. Abstracts: First Chim. Gen. Inorg.*, Milan, Italy. Congr. Naz. Chim. Inorg., [Atti], 12th **1979,** 426-9.

73.   "Transition States: A Preface" Blom, R.; Follestad, A.; Rytter, E.; Tilset, M.; Ystenes, M. <u>Organometallic Catalysts and Olefin Polymerization: Catalysts for a New Millennium</u>. Springer: New York, **2001.** p.V.

74.   For examples of these new compounds, see 58, 59 and the references therein.

187

75.   "Ansa-Metallocene Derivatives. VII. Synthesis and Crystal Structure of a Chiral ansa-Zirconocene Derivative with Ethylene-bridged Tetrahydroindenyl Ligands" Wild, F. R. W. P.; Wasiucionek, M.; Huttner, G.; Brintzinger, H. H. Konstanz, K. *J. Organomet. Chem.*

1985, 288(1), 63-7.

76.   (a) "Homogeneous and High-activity Ziegler-Natta Catalysis with Aluminoxane as Component" Kaminsky, W.; Sinn, H. *Chem. Abstracts: abt. Angew. Chem.* Univ. Hamburg, Fed. Rep. Ger. Proc. IUPAC Macromol. Symp. , 28th **1982**, 247. (b) Kamisky, Walter. *Chem. Abstracts: Inst. Anorg. Angew. Chem.*, Univ. Hamburg, Hamburg, Fed.

Rep. Ger. MMI Press Symp. Ser. **1983**, 4(Transition Met. Catal. Poly.: Alkens Dienes, Pt. A). 225-44. (c) "Olefin Polymerization with Highly Active Soluble Zirconium Compounds Using Aluminoxane as Cocatalyst" Kaminsky, W.; Kuelper, K.; Niedoba, S. *Chem. Abstracts: Makromol. Chem.*, Macromol. Symp. **1986**, 3, 377-87. "Stereoselective

Polymerization of Olefins with Homogeneous Chiral Ziegler-Natta Catalysts" Kaminsky, W. *Chem. Abstracts: Angew. Makromol. Chem.* **1986**, 145-146 149-60.

77.   (a) "Synthesis of Isotactic Polypropylene in a Bubble Column with a Homogeneous Ziegler-Natta Catalyst" Droegemueller, H.; Niedoba, S.; Kaminsky, W. *Chem. Abstracts: Inst. Tech. Makromol. Chem.* Univ. Hamburg, Hamburg, Fed. Rep. Ger. Eds.: Reichert, K.H.; Geiseler, W. Polym. React. Eng,: Emulsion Polym, High Convers. Polym. Polycondens., [Proc. Berlin Int. Workshop], 2nd **1986**, 299-306.

188

(b) "Polymerization of Olefins with Homogeneous Zirconocene/Aluminoxane Catalysts" Kaminsky, W. Steiger, R. *Chem. Abstracts: Polyhedron* **1988**, 7(22-23), 2375-81.

78.   Kaminsky, W.; Sinn, H. "Preface" Transition Metals and Organometallics as Catalysts for Olefin Polymerization: Proceedings of an International Symposium. Eds.: Kaminsky, W.; Sinn, H. Springer-Verlag: New York, **1988**.

79.   Bhaduri, Sumit and Doble, Mukesh. Homogeneous Catalysis: Mechanisms and Industrial Applications. Wiley-Interscience: New York, **2000**. p109-124.

80.   Bhaduri, Sumit and Doble, Mukesh. Homogeneous Catalysis: Mechanisms and Industrial Applications. Wiley-Interscience: New York, **2000**. p5.

81.   Bhaduri, Sumit and Doble, Mukesh. Homogeneous Catalysis: Mechanisms and Industrial Applications. Wiley-Interscience: New York, **2000**. p8.

82.   Boor, John Jr. Ziegler-Natta Catalysts and Polymerizations. Academic Press: New York, **1979**, 393-463.

83.   Blom, R.; Follestad, A.; Rytter, E.; Tilset, M.; Ystenes, M. Organometallic Catalysts and Olefin Polymerization: Catalysts for a New Millennium. Springer: New York, **2001**.

84.   (a) "Stereo- and Enantioselective Polymerization of Olefins with Homogeneous Ziegler-Natta Catalysts" Miller, S. A.; Waymouth, R. M. Eds. Fink, G.; Muelhaupt, R.; Brintzinger, H. H. *Chem. Abstracts: Ziegler Catal.* **1995**, 441-54. (b) "Manipulation of the Ligand Structure as an Effective and Versatile Tool for the

189

Modification of Active Site Properties in Homogeneous Ziegler-Natta Catalyst Systems" Razavi,A.; Vereecke, D.; Peters, L.; Dauw, K. D.; Nafpliotis, L.; Atwood, J. L. Eds. Fink, G.; Muelhaupt, R.; Brintzinger, H. H. *Ziegler Catal.* **1995**, 111-47.

85.   "Cyclohexyl[trans-1,2-bis(1-indenyl)zirkonium(IV)dichlorid: Ein chiraler Polymerisationkatalysator mit stereochemisch starrer Brüücke" Rieger, B. J. *J. Organomet. Chem.* **1992**, 428, C33-36.

86.   "Synthesis and Reactivity of Substituted Metallocenes" Halterman, R. L. Grant Proposal. **1988**.

87.   "Synthesis and Structure of C2-Symmetric, Doubly Bridged Bis(indenyl)titanium and Zirconium Dichlorides" Halterman, R. L.; Tretyakov, A.; Combs, D.; Chang, J.; Khan, M. *Organomet.* **1997**, 16, 3333-3339.

88.   "Synthesis of C7, C7'-Ethylene and C7, C7'-Methylene-Bridged C2-Symmetric Bis(indenyl)zirconium and -titanium Dichlorides" Halterman, R. L.; Combs, D.; Khan, M. A. *Organomet.* **1998**, 17, 3900-3907.

89.   (a) "Stereochemistry at the Migration Terminus in the Base-induced Rearrrangement of alpha-haloorganoboranes" Midland, M. M.; Zolopa, A R.; Halterman, R. L.. *Chem. Abstracts: J. Am. Chem. Soc* **1979**, 01(1), 248-9. (b) "Stereoselective Bromination of Allylic Alcohols. A Facile Synthesis of (E)- or (Z)-bromoepoxides from a Comon Starting Material" Midland, M. M.; Halterman, R. L. *Chem. Abstracts: J. Org. Chem.*

190

1981, 46(6) 1227-9.

90. "The Isomerization of Optically-active Propargyl alcohols to Terminal Acetylenes" idland, M. M.; Halterman, R. L.; Brown, C. A.; Yamaichi, A. . *Chem. Abstracts: Tet. Lett.* **1981**, 22(42), 4171-2.

91. (a) "Cobalt-Mediated Synthesis of Tricyclic Dienes Incorporating Fused Four-Member Rings. Unprecedented rearrangements and Structural Characterization of a Cobalt-diene Complex by Two-dimensional NMR Spectroscopy" Dunach, E.; Halterman, R. L.; Vollhardt,, K. P. C. . *Chem. Abstracts: J. Am. Chem. Soc.* **1985**, 107(6), 1664-71. (b)"Steric Hindrance to Benzcyclobutene Openings. First Synthesis of a 1,2,3-tris(trimethylsilated) arene by Cobalt-catalyzed Alkyne Cyclizations and Application of Fully Coupled Two-Dimensional Chemical Shift Correlations to a Structural Problem" Halterman, R. L.; Nguyen, N. H.; Vollhardt, K. P. C. *Chem. Abstracts: J. Am. Chem. Soc.*
1985, 107(5), 1379-87.

92. "Syntheis and Asymmetric Reactivity of Electronically and Sterically Differentiating Chiral Cyclopentadienyl Metal Complexes" Halterman, R. L. *Chem. Abstracts: Diss. Abstr. Int. B* **1986**, 46(11), 3846-7.

93. "Diisopropyl Tartrate Modified (E)-crotylboronates: Highly Enantioselective Propionate (E)-enolate Equivalents" Roush, W. R.; Halterman, R. L. *Chem. Abstracts: J. Am. Chem. Soc.* **1986**, 108(2), 294-6.

94. "Binapthylcyclopentadiene: a C2-symmetric Annulated Cyclopentadienyl Ligand

191

with Axial Chirality" Colletti, S. L.; Halterman, R. L. *Tet. Lett.* **1989**, 30(27), 3513-16.

95.   Combs, David. <u>The Synthesis of Group IV Metallocene Dichlorides with Variable Sterics, Conformational Mobility and Geometric Shapes.</u> Ph.D. Dissertation, **1997**.

96.   Tretyakov, Alexander. <u>New Strategies for the Synthesis of Cyclopentadienyl and Indenyl Metallocene Dichloride Complexes.</u> Ph.D. Dissertation, **1995**.

97.   Ramsey, Timothy Michael. <u>Design and Synthesis of Electronically and Sterically Variable Chiral Cyclopentadienyl Transition Metal Complexes and their Application in Asymmetric Transformations.</u> Ph.D. Dissertation, **1994**.

98.   (a) "α-Ketosääuren als ÄÄquivalent füür Aldehyde in der Thaizloium-salz-katalysierten Additin" Stetter, H.; Lorenz, G. *Chem. Ber.* **1985**, 118, 1115-1125. (b) "Addition of aldehydes to activated double bonds. VI. Addition of aliphatic aldehyde to methyl vinyl ketone" Stetter, H.; Kuhlmann, H. Hochsch, A. *Tet. Lett.* **1975** 169947. (c) "Addition of aldehydes to activated double bonds. 5. Addition of aliphatic aldehydes to activated double bonds" Stetter, H.; Kuhlmann, H.; Hochsch, A. *Angewandte Chemie* **1974**, 86(16), 589. (d) "Catalytic oxidation of carbonyl compounds" Goldstein, T.P. *US Patent* AN **1974**: 449438. (e) "A new catalyst for acyloin condensation. I. " Ukai, T.; Tanaka, R.; Dokawa, T. *AN* **1951**: 29649.

99.   "A Convenient Metod for the Synthesis of Insect Growth Regulators: Cyclopentene

Analogs of Alkyl (2E, 4E) Dodecadienoates" Novåk, L.; Rohåly, J.; Gålik, G.; Fekete, J.; Varjas, L.; SzEantay, C. Liebigs *Ann. Chem.* **1986**, 509-524.

100.   "Preparation of Substituted Indenes by Cyclocondensation of Alkanediones with Cycopentadienes" Erker, G.; Nolte, R.; Aulbach, M.; Weiss, A.; Reuschling, D.; Rohrmann, *J. Ger. Offen.* **1992**, Patent.

101.   (a) "Synthesis, Structure and Properties of Chiral Titanium and Zirconium Complexes Bearing Biaryl-strapped Substituted Cyclopentadienyl Ligands" Ellis, W. W.; Hollis, K. T. ; Udenkirk, W.; Whelan, J.; Ostrander, R.; Rheingold, A. L.; Bosnich, B. *Organomet.* **1993**, 4391-4401. (b) "Chemistry of O-Xylidene-Metal Complexes. Part 1. O-Xylidene- Magnesium Reagents as Metallocyclic Precursors and Synthesis of [Pt(CH2C6H4Ch2-O)(cod)] (cod=cyclo-octa-1,5-diene); the X-Ray Crystal Structure of the Macrocycle [[Mg(CH2C6H4CH2-O)(C4H8O)2]3]" Lappert, M. F.; Martin, T. R.; Raston, C. L.; Skeleton, B. W.; White, A. H. *J. Chem. Soc, DT* **1982**, 1959.

102.   "Efficient Synthesis of rac-(Ethylenebis(indenyl))ZrX$_2$ Complexes via Amine Elimination" Diamond, G. M.; Rodewald, S.; Jordan, R. F. *Organomet.* **1995**, 5-7.

103.   "Witting reaction on carbonyl-containing cyclotriphosphazenes: the case of hexakis(4-formylphenoxy)cyclophosphazene" Facchin, G.; Bertani, R.; Gleria, A. B. *Inorganic Chemica Acta*, **1988**, 147, 165-171.

104.   a) "Synthesis of new C2-symmetrical diphosphines using chiral zinc organometallics". Longeau, A.; Durand, S;; Spiegel, A; Knochel, Pl.

193

Tetrahedron: Asymmetry **1997**, 8(7), 987-990. b) "Brominative decarboxylation of optically active silver trans-1,2-cyclohexanedicarboxylate". Applequist, D. E.; Werner, N D. *J. Org. Chem.* **1963**, 28 48-54.

105. (a) "Formation of α,β-unsaturated carboxylic acid derivatives by the Wittig reaction." Takamatsu, Hideji; Umemoto, Susumu; Shimizu, Toyoyuki; Kagemoto, Akira. *Yakugaku Zasshi* **1965**, 85(11), 975-80. (b) "The Wittig reaction with five- and six-membered cyclic ketones and their benzylidene derivatives." Witschard, G.; Griffin, C. E. *J. Org. Chem.* **1964**, 29(8), 2335-40. (c) "Photolysis of 1,1,1-triarylalk-2-enes and 1,1,1-triarylhept-2-ynes. A novel generation of aryl-1-alkenyl and aryl-1-alkynyl arbenes." Shi, Min; Shouki, Kouji; Okamoto, Yoshiki; Takamuku, Setsuo. *J. Chem. Soc., PT1: Organic and Bio-Organic Chemistry (1972-1999)* **1990**, (9), 2443-50. (d) "Butenolide synthesis. Part 3. Convenient synthesis of 5-alkylidene-2(5H)-furanones, 2(5H)-furanones, and 2-ethoxyfurans." Saalfrank, R. W.; Hafner, W.; Markmann, J.; Bestmann, H. J. *Tetrahedron* **1988**, 44(16), 5095-100. (e) "Butenolide syntheses. II. Simple synthesis of 4-substituted 2,2-diethoxy-5-alkylidene-2,5-dihydrofurans, 3-substituted 4-alkylidene-2-buten-4-olides and 5,6-dihydro-2-pyrones." Saalfrank, Rolf W.; Schierling, Peter; Hafner, Wieland. *Chem. Ber.* **1983**, 116(10), 3482-6. (f) "Capsaicinoids: their separation, synthesis, and mutagenicity." Gannett, Peter M.; Nagel, Donald L.; Reilly, Pam J.; Lawson, Terence; Sharpe, Jody; Toth, Bela. *J. Org. Chem..* **1988**, 53(5), 1064-71.

106. (a) "(E)-Selective Horner-Wadsworth-Emmons reaction of aryl alkyl ketones with

194

bis(2,2,2-trifluoroethyl)phosphonoacetic acid." Sano, Shigeki; Takemoto, Yuka; Nagao, Yoshimitsu. *Tetrahedron Letters* **2003**, 44(49), 8853-8855. (b) "Development of highly stereoselective reactions utilizing heteroatoms--new approach to the stereoselective Horner-Wadsworth-Emmons reaction." Sano, S. *Yakugaku zasshi. Journal of the Pharmaceutical Society of Japan* **2000** 120(5), 432-44. (c) "Stereocontrolled Total Synthesis of 1α,25-Dihydroxycholecalciferol and 1α,25-Dihydroxyergocaliferol" Baggiolini, E. C.; Iacobelli, J. A.; Hennessy, B. M.; Batcho, A. D.; Sereno, J. F.; Uskokovic, M. R. *J. Org. Chem.* **1986**, 3098-3108.

107.   (a) "Friedel-Crafts acylation." Metivier, Pascal. *Fine Chemicals through Heterogeneous Catalysis* **2001**, 161-172 (b) "Friedel-Crafts acylation: Interactions between Lewis acids-acyl chlorides and Lewis acids-aryl ketones." Ashforth, Rebecca; Desmurs, Jean-Roger. Industrial Chemistry Library **1996**, 8(Roots of Organic Development), 3-14. (c) "Friedel-Crafts acylations with little or no catalyst." Pearson, D. E.; Buehler, Calvin A. *Synthesis* 1972, (10), 533-42.

108.   "Oxidative Coupling of the Enolate Anion of (1R)-(+)-Verbenone with Fe(III) and Cu(II) Salts. Two Modes of Conjoining This Bicyclic Ketone across a Benzene Ring". Paquette, L.A.; Bzowej, E. I.; Branan, B. M.; Stanton, K. J. *J. Org. Chem.* **1995**, 60(22), 7277-83.

109.   "Efficient synthesis of 2- and 3-substituted indenes from 2-bromobenzyl bromide through an enolate alkylation/Cr(II)/Ni(II)-mediated carbonyl addition sequence".

195

Halterman, R. L.; Zhu, C. *Tet. Lett.* **1999**, 40(42), 110.

110.   a) Zhu and Halterman unpublished results, b) "Synthesis of 1,4-diketones by oxidative coupling of ketone enolates with copper(II) chloride" Ito, Y.; Konoike, T.; Harada, T.; Saegusa, T. *J. Am. Chem. Soc.* **1977**, 1487-93. (c) "Oxidative coupling of ketone enolates by ferric chloride" Frazier, R. H., Jr.; Harlow, R. L. *J. Org. Chem.* **1980**, 5408.

111.   "Molecular meccano. Part 47. [C-H·····O] Interactions as a control element in supramolecular complexes: experimental and theoretical evaluation of receptor affinities for the binding of bipyridinium-based guests by catenated hosts" Houk, K. N.; Menzer,S.; Newton, S.P.; Raymo, F.M.; Stoddart, J.F.; Williams D. J. *J. Am. Chem. Soc.* **1999**, 121(5), 1479-1487.

112.   (a) "(S)-(-)-1-amino-2-methoxymethylpyrrolidine (SAMP) and (R)-(+)-1-amino-2-methoxymethylpyrrolidine (RAMP) as versatile chiral auxiliaries" Mujahid, A. M. *Synlett* **2003**, (13), 2099-2100. (b) "Asymmetric b-aminoethylation of ketones and nitriles with tosylaziridines employing the SAMP-hydrazone method" Enders, D; Janeck, C. F.; Raabe, G. *European Journal of Organic Chemistry* **2000**, 19, 3337-3345. (c) "Asymmetric synthesis of b-substituted d-keto esters via Michael additions of (S)-AMP/(R)-AMP hydrazones [AMP = 1-amino-2-methoxymethylpyrrolidine] to a,b-unsaturated esters. Virtually complete 1,6-asymmetric induction" Enders, D.; Papadopoulos, K. *Tetrahedron Lett.* **1983**, 24(45), 4967-70. (d) "Asymmetric synthesis of both enantiomers of (E)-4,6-dimethyl-6-octen-3-one, the defensive substance of daddy longlegs, Leiobunum vittatum and L. calcar (Opiliones)" Enders, D.; Baus, U. *Liebigs Annalen der*

196

Chemie **1983**, (8), 1439-45.

113.   (a) "Regiospecific synthesis of aza-b-lactams from diaziridines" Alper, H.; Delledonne, D.; Kameyama, M.; Roberto, D. *Organometallics* **1990**, 9(3), 762-5 (b) "Transannular cyclization as a stratagem in synthesis. Total synthesis of (±)-pentalenene" Pattenden, G.; Teague, S. J. *Tetrahedron* **1987**, 43(23), 5637-52. (c) "3-Hydroxypyrroles and 1H-pyrrol-3(2H)-ones. Part 1. Formation and x-ray crystal and molecular structure of 1-cyclohexyl-2,2-pentamethylene-1H-pyrrol-3(2H)-one" Hickson, C. L.; Keith, E.M.; Martin, J. C.; McNab, H.; Monahan, L. C.; Walkinshaw, M. D. *J. Chem. Soc., PT1: Organic and Bio-Organic Chemistry (1972-1999)* **1986**. 8, 1465-9. (d) "Alcohols and aluminum alkoxides in the presence of Raney nickel as alkylating agents. 3. Reduction of Schiff bases with isopropyl alcohol and aluminum isopropoxide in the presence of Raney nickel" Lotta, M.; De Angelis, F.; Gambacorta, A.; Labbiento, L.; Nicoletti, R. *J. Org. Chem.* **1985**, 50(11), 1916-19.

114.   (a) Halterman and Zhu. Unpublished Results. (b) "Kinetics and mechanisms of the oxidation of methyl aryl ketones by acid permanganate" Marigangaiah; N., Permashwar, B., Kalyan K. *Australian Journal of Chemistry* **1976**, 29(9), 1939-45.

115.   a) "Oxidative Coupling of the Enolate Anion of (1R)-(+)-Verbenone with Fe(III) and Cu(II) Salts. Two modes of Conjoining This Biyclic Ketone across a Benzene Ring". Paquette, L. A.; Bzowej, E. I.; Branan, B. M.; Stanton, K. J. *J. Org. Chem.* **1955**, 60(22), 7277-83. AN: 1995:865233 CAPLUS b) "Copper catalyzed coupling

197

reactions". Lipshutz, B.H. *PCT Int. Appl.* **1994**, AN 1995:346836 CAPLUS

116.   "Synthesis of 1,4-diketones by reductive coupling of alpha,alpha'-
dibromoketones". Chassin, C.; Schmidt, E.A.; Hoffman, H. M. R. *J. Am. Chem.
Soc.* **1974**, 96(2), 606-8. AN 1974:81990 CAPLUS.

117.   (a) "New reactivity of functionalized organolithium compounds in the presence of
Cu(I) or Cu(II) salts: conjugate addition, acylation or dimerization" Pastor, I. M.;
Yus, M.   *Tetrahedron* **2001**, 57(12), 2371-2378. (b) Pastor, Isidro M.; Yus,
Miguel. "Copper(I) or (II)-mediated conjugate addition or dimerization of
functionalized organolithium compounds" *Tetrahedron Let.* **2000**, 41(10),
1589-1592. (c) "Selective Copper-Catalyzed Coupling Reactions of (a-
Acetoxyhexyl)tricyclohexyltin" Linderman, R.J.; Siedlecki, J. M. *J. Org. Chem.*
1996, 61(19), 6492-6493. (d) "Copper-catalyzed oxidation of unsaturated carbonyl
compounds. I. Catalyzed oxidation of 5-cholestenone to 4-cholestene-3,6-dione"
Volger, H. C.; Brackman, W. *Recueil des Travaux Chimiques des Pays-Bas"*
1965, 84(4), 579-80.

118.   Organic Syntheses: Collective Volume II: A Revised Edition of Annual Volumes
X-XIX. Blatt, A.H. ed.; J. Wiley & Sons, Inc. New York: 1943. p. 569.

119.   Vogel's Textbook of Practical Organic Chemistry, 5th ed., Farniss, B.S.;
Hannaford, A.J.; Smith, P.W.G.; Tatchell, A.R.; eds. Longman Scientific &
Technical. New York, 1989, p.608.

120.   "Indenyl-and Fluorenylsilanes: Synthesis and Thermal Diatereomerization,"
Raushc, M. D.; Chen, Y.; Chien, J. *Organometallics* **1993**, 12, 4607-12.

121. "The Preparation, Separation, and Characterization of the lel$_3$ - and ob$_3$- Isomers of Tris (*trans*-1,2-cyclohexanediamine) rhodium (III) Complexes" Galsbol, F.; Steenbol, P.; Sorensen, B.S.; *Acta Chem. Scand* **1972**, 26, 3605-3611.

122. "Catalytic Asymmetric Addition of Polyfunctional Dialkylzines to β-Stannylated and β-Silylated Unsaturated Aldehydes." Ostwald, R.; Chavant, P.; Stadtmuller, Y.; Knochel, P. *J. Org. Chem.* **1994**, 59, 15, 4148.

123. (a) "Indenyl- and Fluorenylsilanes: Synthesis and Thermal Diastereomerization." Chien,C.W.J.; Rausch, M.D.; Chen, Y. X. *Organometallics* **1993**, 12, 4608 (b)"Di Indenyl Magnesium" Smith, K. D.; Atwood, J. L. Inorganic Syntheses. Volume XVI. Annual Reports in Inorganic and General Chemistry" ed. Fred Basolo. Academic Press: New York, 1976, p137.

124. "Heptane-Soluble Homogeneous Zirconocene Catalyst: Synthesis of a Single Diastereomer, Polymerization Catalysis, and Effect of Silica Supports" Chen, Y.X.,; Rausch, M.D.; Chien, J.C.W. *J. of Poly. Sci. Part A: Poly. Chem.* **1995**, 3(33), 2093-2108.

125. "Catalytic Asymmetric Addition of Polyfunctional Dialkylzines to β-Stannylated and β-Silylated Unsaturated Aldehydes." Ostwald, R.; Chavant, P.; Stadtmuller, Y.; Knochel, P. *J. Org. Chem.* **1994**, 59(15), 4143-4153.

126. (a) "Enantioselective Michael reactions. Stereoselective addition of enolates of phenmenthol esters to crotonates" Corey, E. J.; Peterson, R. T. *Tetrahedron Let.* **1985**, 26(41), 5025-8. (b) "Highly enantioselective Michael reactions catalyzed by a chiral quaternary ammonium salt. Illustration by asymmetric syntheses of (S)-

199

ornithine and chiral 2-cyclohexenones" Zhang, F.; Corey, E. J. *Organic Letters* **2000**, 2(8), 1097-1100. (c) "Enantio- and diastereoselective michael reactions of silyl enol ethers and chalcones by catalysis using a chiral quaternary ammonium salt" Zhang F.; Corey, E. J. *Organic Letters* (**2001** Feb 22), 3(4), 639-41. (d) "Enantioselective Michael addition of nitromethane to a,b-enones catalyzed by chiral quaternary ammonium salts. A simple synthesis of (R)-baclofen" Zhang F.; Corey, E. J. *Organic Letters* **2000,** 2(26), 4257-4259.

127.   (a) "Synthesis, Characterization, and Polymerization Properties of Bis(2-menthylindenyl)zirconium Dichloride and Bis(2-menthyl-4,7-dimethylindenyl)zirconium Dichloride" Halterman, R.L.; Fahey, D. R.; Bailly, E. F.; Dockter, D.W.; Stenzel, O.; Shipman, J. L.; Khan, M.A.; Dechert, S.; Schumann, H. *Organometallics* **2000**, 19(25), 5464-5470. (b) "Synthesis and characterization of 1- and 2-(w-alken-1-yl)indenes, their lithium salts and dichlorozirconium(IV) complexes" Schumann, H.; Karasiak, D. F.; Muhle, S.H.; Halterman, R. L.; Kaminsky, W; Weingarten, U. *Journal of Organometallic Chemistry* **1999**, 579(1-2), 356-372. (c) "Synthesis and structure of [1,2-bis(1-indenyl)benzene]titanium and zirconium dichlorides" Halterman, R. L.; Tretyakov, A.r; Khan, M.A. *Journal of Organometallic Chemistry* **1998,** 568(1-2), 41-51. (d) "Synthesis of ansa-2,2'-bis[(4,7-dimethylinden-1-yl)methyl]-1,1'-binaphthyl and ansa-2,2'-bis[(4,5,6,7-tetrahydroinden-1-yl)methyl]-1,1'-binaphthyltitanium and -zirconium dichlorides") Halterman, R. L.; Combs, D.; Kihega, J.; Khan, M.A. *Journal of Organometallic Chemistry* **1996**, 520(1-2),

200

163-170.

128.   "Analgesics: Synthesis of 1-Dioxolanoalkylnormeperidines and 1-Diaxanoalkylnormeperidines" Kriesel, D.C.,; Gisvold, O. *J. Pharm Sci* **1971**, 60 1250-1251.

129.   (a)"Synthesis of epi-β-Santalene, β-Santalene and an isomer of β-Santalene with 4-Methyl-4-Pentenyl Side Chain" Unnikrishnan, P.A.; Vatakencherry, P.A. *Syn. Comm.* **1992**, 22 (22) 3159-3168 (b)"3,3-Ethylenedioxybutyl-magnesiumbromide-A Nucleophilic 3-Ketobutyl Equivalent" Ponaras, A. A. *Tet. Lett.* **1976**, 36, 3105-3108. (c) "A New Synthesis of 1,4-Diketones: Application to a Synthesis of Dihydrojasmone and *cis*-Jasmone" Faculty of Pharmaceutical Sciences, Hokkaido University, Sapporo, Japan. *Tet. Lett.* **1974**, 44, 3883-3886.

130.   Note: While Vladimir claims the technique is common, I have not seen it specifically set out in any literature to date: For each gram of LAH, quench with 1.00 mL of water, add 1.00 mL of 1M NaOH, cease stirring, compact solid with 3.00 mL of water, add and decant ether from aluminum hydroxide precipitates.

131.   (a) "Synthesis, Structure, and Properties of Chiral Titanium and Zirconium Complexes Bearing Biaryl Strapped Substituted Cyclopentadienyl Ligands." Ellis, W.W.; Hollis, K.T.; Odenkirk, W.; Whelan, J.; Ostrander, R.; Rheingold, A. L.; Bosnich, B. *Organometallics* **1993**, 12, 4391-4401 (b) "Chemistry of o-Xylidene-Metal Complexes. Part 1.o-Xylidene-magnesium Reagnets as Metallocyclic Precursors and Synthesis of {Pt(CH$_2$C6H4,CH$_2$-o) (cod)] (cod=cyclo-octa-1,5-diene); the X-Ray Crystal Structure of the Macrometallocycle [{Mg(CH$_2$C$_6$H$_4$CH$_2$-

201

o) (C$_4$H$_8$O)$_2$}$_3$" Lappert, M.F.; Martin, T.R.; Raston, C. L.; Skelton, B.W.; White, A.H. *J. Chem. Soc, D.T.* **1982**, 1959-1964.

132.   (a) "Michael addition of masked thiolates to conjugated systems in aqueous media promoted by ammonium tetrathiomolybdate" Devan, N.; Sureshkumar, D.; Beadham, I.; Prabhu, K. R.; Chandrasekaran, S. *Indian Journal of Chemistry, Section B: Organic Chemistry Including Medicinal Chemistry* **2002**, 41B(10), 2112-2115. (b) "Synthesis of novel bicycloalkane ring-fused pyridines via Michael addition reaction of a,b-unsaturated nitriles" El-Shehawy, A. A.; Ismail, A. A.; Attia, A. M. *Sulfur Letters* **2003**, 26(2), 55-62. (c) "Triphenylphosphine catalyzed Michael addition of oximes onto activated olefins" Bhuniya, D.; Mohan, S.; Narayanan, S. *Synthesis* **2003**, (7), 1018-1024. (d) "Synthesis of a Heteroannulated Furanose via Intramolecular Michael Addition in an w-Hydroxy a,b-Unsaturated Nitrile" Marco, J. L.; Fernandez, S.. *Journal of Chemical Research, Synopses* **1999**, (9), 544-545. (e) "Diverse Cycloaddition Chemistry Leading to Overall Michael Addition in the Reactions of 1,1-Bis(dimethylamino)-2,2-difluoroethene with α,β-Unsaturated Aldehydes, Ketones, Esters, and Nitriles" Xu, Y.; Dolbier, W.R., Jr. *Journal of Organic Chemistry* **1997**, 62(19), 6503-6506.

133.   (a) "Synthesis of functionalized cyclic imines by addition of Grignard reagents to ω-bromonitriles and α,β-Unsaturated nitriles" Fry, D. F.; Brown, M.; McDonald, J. C.; Dieter, R. K. *Tetrahedron Letters* **1996**, 37(35), 6227-6230. (b) "Grignard reagent additions to α,β-ethylenic nitriles: 1,4- vs. 1,2- additions" Kulp, S. S.;

202

DiConcetto, J. A. *Journal of Chemical Education* **1990**, 67(3), 271-3. (c) "The reaction of organo-magnesium compounds on nitriles. Glutaronitrile" Bruylants, P. *Bulletin des Societes Chimiques Belges* **1923**, 32, 307. (d) "Action of the Grignard reagent upon amino nitriles" Stevens, T.S.; Cowan, J. M.; MacKinnon, J. *Journal of the Chemical Society, Abstracts* **1931**, 2568-72.

134.   (a) "A novel reduction of nitriles to aldehydes" Backeberg, O. G.; Staskun, B. *Journal of the Chemical Society, Abstracts* **1962**, 3961-3. (b) "Reductions with Raney alloy in acid solution" van Es, T.; Staskun, B. *Journal of the Chemical Society, Abstracts* **1965**, 5775-7. (c) "Reductive hydrolysis of nitriles to aldehydes by precipitated nickel on iron or aluminum" Bodnar, Z.; Mallat, T.; Petro, J. *Journal of Molecular Catalysis* **1991**, 70(1), 53-64. (d) "One-pot transformation of nitriles into aldehyde tosylhydrazones" Toth, M.; Somsak, L. *Tet. Letters* **2001**, 42(14), 2723-2725. (e) "Transfer Hydrogenation of Nitriles with 2-Propanol and Raney Nickel Mebane" Robert C.; Jensen, D. R.; Rickerd, K. R.; Gross, B. H.. *Synthetic Communications* **2003**, 33(19), 3373-3379.

135.   "The anomalous Hunsdiecker reaction of triaryl-substituted aliphatic acids" Wilt, J. W.; Lundquist, J.A. *J. Org. Chem.* **1964**, 29(4), 921-5.

136.   "Modifications of the Hunsdiecker reaction" Davis, J. A.; Herynk, J.; Carroll, S.; Bunds, J.; Johnson, D. *J. Org. Chem.* **1965**, 30(2), 415-17.

137.   "Synthesis of Brexan-2-one and Ring-Expanded Gongeners" Stern, A.G.; Nickon, A.; Trost, B.M.; *J. Org. Chem.* **1992**, 57, 20, 5342-5352.

138.   (a) "Synthesis, Characterization, and Polymerization Properties of Bis(2-

203

menthylindenyl)Zirconium Dichloride and Bis (2-menthyl-4,7-dimethylindenyl) zirconium dichloride" Halterman, R.L.; Fahey, D. R.; Bailly, E.F.; Dockter, D. W.; Stenzel, O.; Shipman, J. L.; Khan, M. A. ; Dechert, S.; Schumann, H. *Organomet.* 2000, 19(25), 5469. (b) "Titanium + Zirconium Complexes of 2-Methyindone" Shipman, J. L., M.S. Thesis, University of Oklahoma 1997. p. 35.

139. (a) "Titanium + Zirconium Complexes of 2-Methyindone" Shipman, J. L., M.S. Thesis, University of Oklahoma 1997. p. 27, (b) "Synthesis of 2-substituted Indenes Using a Nickel Catalyzed Cross-Coupling Technique" Seljeseth, C. Senior Thesis University of Oklahoma 1997 (c)"Synthesis of Biological Markers in Fossil Fuels. 2. Synthesis and ^{13}C NMR Studies of Substituted Indans and Tetralins" Adamcyzk, M.; Watt, D.S.; Netzel, D. A. *J. Org. Chem* 1984, 49(22), 4226-4237. (d) "Synthesis, Characterization, and Polymerization Properties of Bis(2-menthylindenyl)Zirconium Dichloride and Bis (2-menthyl-4,7-dimethylindenyl) zirconium dichloride" Halterman, R.L.; Fahey, D. R.; Bailly, E.F.; Dockter, D. W.; Stenzel, O.; Shipman, J. L.; Khan, M. A. ; Dechert, S.; Schumann, H. *Organomet.* 2000, 19(25), 5469. (e) "Syntheses of [Ethylene-1,2-bis(η^5-4,5,6,7-tetrahydro-1-indenyl)] zirconium and -hafnium hydride complexes. Improved Syntheses of the Corresponding Dichlorides." Grossman, R.B.; Doyle. R. A.; Buchwalk, S.L.; *Organomet.* 1991 (10) 1501-5. (f)"Synthesis and Characterization of (-) -1-Menthyl-4,7-dimethylindene and its Main Group Metal Compounds with Lithium, Sodium, Potassium and Tin" Schumann, H.; Stenzl, O.; Girgsdies, F.; Halterman, R.L. *Organomet.* 2001, 1744.

204

140. "Syntheses of [Ethylene-1,2-bis(η⁵-4,5,6,7-tetrahydro-1-indenyl)] zirconium and -hafnium hydride complexes. Improved Syntheses of the Corresponding Dichlorides." Grossman, R.B.; Doyle. R. A.; Buchwalk, S.L.; *Organomet.* **1991** (10) 1501-5.

141. "Nitrodibromoacetonitrile: an agent for bromination and for the formation of adducts formally derived from cyanonitrocarbene" Boyer, J. H.; Manimaran, T. *J. Chem. Soc., P.T. 1: Organic and Bio-Organic Chemistry (1972-1999)* **1989**, (8), 1381-5.

142. "The Diastereomeric Menthylchlorides Obtained from (-)-menthol" Smith, J. G.; Wright, G. F. *J. Org. Chem.* **1952**, 17, 1116-1120.

143. "Titanium and Zirconium Complexes of 2-Methyindone" Shipman, J. L., M.S. Thesis, University of Oklahoma **1997**. p. 27.

144. (a) "Synthesis, Characterization, and Polymerization Properties of Bis(2-menthylindenyl)zirconium Dichloride and Bis(2-menthyl-4,7-dimethylindenyl)zirconium Dichloride." Halterman, R.L.; Fahey, D. R.; Bailly, E. F.; Dockter, D. W.; Stenzel, O.; Shipman, J. L.; Khan, M. A.; Dechert, S.; Schumann, H. *Organometallics* **2000**, 19(25), 5464-5470. (b) "Synthesis, structure, and catalytic activity of some new chiral 2-menthylindenyl and 2-menthyl-4,7-dimethylindenyl rhodium complexes." Schumann, H.; Stenzel, O.; Dechert, S.; Girgsdies, F.; Blum, J.; Gelman, D.; Halterman, R. L.. *European Journal of Inorganic Chemistry* **2002**, (1), 211-219.(c) "(-)-2-Menthylindenyl and (-)-2-Menthyl-4,7-dimethylindenyl Complexes of Rhodium, Iridium, Cobalt,

205

and Molybdenum." Schumann, H.; Stenzel, O.; Dechert, S.; Girgsdies, F.; Halterman, R. L.. *Organometallics* **2001**, 20(25), 5360-5368. (d)"Menthyl-Functionalized Chiral Nonracemic Monoindenyl Complexes of Rhodium, Iridium, Cobalt, and Molybdenum." Schumann, H.; Stenzel, O.; Dechert, S.; Girgsdies, F.; Halterman, R. L. *Organometallics* **2001**, 20(11), 2215-2225. (e) "Menthyl-Functionalized Chiral Nonracemic Bis(indenyl) Complexes of Zirconium, Iron, Nickel, and Ruthenium." Schumann, H.; Stenzel, O.; Dechert, S.; Halterman, R. L. *Organometallics* **2001**, 20(10), 1983-1991. (f)"Synthesis and Characterization of (-)-1-Menthyl-4,7-dimethylindene and Its Main Group Metal Compounds with Lithium, Sodium, Potassium, and Tin". Schumann, H.; Stenzel, O.; Girgsdies, F.; Halterman, R. L.. *Organometallics* **2001**, 20(9), 1743-1751.

145. For relatively recent reviews/information on Ring Closing Metathesis see: (a) "Catalytic ring-closing metathesis and the development of enantioselective processes." Hoveyda, A.H. *Topics in Organometallic Chemistry* **1998**, 1(Alkene Metathesis in Organic Synthesis), 105-132. (b) "Olefin metathesis by well-defined complexes of molybdenum and tungsten." Schrock, R. R. *Topics in Organometallic Chemistry* **1998**, 1(Alkene Metathesis in Organic Synthesis), 1-36. (c) "Some recent applications of olefin metathesis in organic synthesis: A review." Ivin, K. J. *J. Mol. Cat. A: Chemical* **1998**, 133(1-2), 1-16. (d) "Applications of Zr-catalyzed carbomagnesation and Mo-catalyzed macrocyclic ring closing metathesis in asymmetric synthesis. Enantioselective synthesis of Sch

206

38516 (fluvirucin B1)." Scheidt, K. A.; Roush, W. R.. *Chemtracts* **1998**, 11(7), 522-530. (e) "Olefin metathesis in organic chemistry." Schuster, M.; Blechert, S. *Angewandte Chemie, International Edition in English* **1997**, 36(19), 2037-2056. (f) "Ring-Closing Metathesis and Related Processes in Organic Synthesis." Grubbs, R. H.; Miller, S. J.; Fu, G. C.*Accounts of Chemical Research* **1995**, 28(11), 446-52.

146.   (a) "Ozonolyzes of 1-Alkyl-Substituted 3-Methylindenes. Remarkable Effects of the Substituent Steric Bulk and the Stereochemistry of the Carbonyl Oxide Intermediates on the Efficiency of Ozonide Formation" Kawamura, S.; Takeuchi, R.; Masuyama, A.; Nojima, M.; McCullough, K. J. *J.of Org. Chem.* **1998**, 63(16), 5617-5622. (b) "Ozonolyses of Indene and of 1-Alkyl- and 1,1-Dialkyl-Substituted Indenes in Protic Solvents. Remarkable Effects of the Substituent Steric Bulk and the Solvent Nucleophilicity on the Direction of Cleavage of the Primary Ozonides and on the Mode of Capture of the Carbonyl Oxide Intermediates by the Solvents") Teshima, K.; Kawamura, S.; Ushigoe, Y.; Nojima, M.; McCullough, K. J. *J. Org. Chem.* **1995**, 60(15), 4755-63. (c) "Ozonolysis of indene in ethanol" Warnell, J. L.; Shriner, R. L. . *J. Am. Chem. Soc.* **1957**, 79 3165-7.

147.   "Synthetic and mechanistic aspects on the Wittig reaction. A yield increasing modification" Adlercreutz, P.; Magnusson, G. *Acta Chem. Scand., B: Organic Chemistry and Biochemistry* **1980**, B34(9), 647-51.

148.   (a) "Reaction of Phenyl-Substituted Allyllithiums with tert-Alkyl Bromides.

207

Remarkable Difference in the Alkylation Regiochemistry between a Polar Process and the One Involving Single-Electron Transfer" Tancks, J.; Nojma, M.; Kusabayashi, S. *J. Am. Chem. Soc.* **1987** 109, 3391-3397. (b) "Reaction of (E)-β-Enamino Amides with Dimethyl Acetylenedicarboxylate (DMAD): Formation of Benzene Derivatives, Enamino Esters, and 2-Pyridones" Nuvole, A.; Paglietti, G. *J. Chem. Soc. P.T. II* **1989**, 1007-1011.

149. (a) "Preparation of 1,1'-binaphthyl-2,2'-diyl bridged ansa-bis(annulated cyclopentadienyl)titanium and -zirconium dichloride complexes." Halterman, R. L.; Ramsey, T. M. *J. Organomet. Chem.* **1997**, 530(1-2), 225-234. (b) "Synthesis of ansa-2,2'-bis[(4,7-dimethylinden-1-yl)methyl]-1,1'-binaphthyl and ansa-2,2'-bis[(4,5,6,7-tetrahydroinden-1-yl)methyl]-1,1'-binaphthyltitanium and -zirconium dichlorides." Halterman, R. L.; Combs, D.; Kihega, J.; Khan, M. A. *J. Organomet. Chem.* **1996**, 520(1-2), 163-170. (c) "Bis(binaphthylcyclopentadienyl)-derived metallocene peroxide complexes: catalysts for the enantioselective epoxidation of unfunctionalized alkenes." Colletti, S. L.; Halterman, R. L. *J. Organomet. Chem.* **1993**, 455(1-2), 99-106. (d) "Asymmetric epoxidation of unfunctionalized alkenes using the new C2-symmetrical 1,1'-binaphthyl-2,2'-dimethylene-bridged ansa-bis(1-indenyl)titanium dichloride catalyst." Colletti, S. L.; Halterman, R. L. *Tet. Lett.* **1992**, 33(8), 1005-8. (e) "Asymmetric synthesis and metalation of a binaphthylcyclo-pentadiene, a C2-symmetric chiral cyclopentadiene." Colletti, S. L.; Halterman, R.L. *Organomet.* **1991**, 10(10), 3438-48. (f) "C2-symmetric 2,2'-dimethyl-1,1'-

208

binaphthyl-bridged ansa-bis(1-indenyl)metal complexes." Burk, M. J.; Colletti, S. L.; Halterman, R. L. *Organomet* **1991**, 10(9), 2998-3000. (g) "Binaphthylcyclopentadiene: a C2-symmetric annulated cyclopentadienyl ligand with axial chirality." Colletti, S. L.; Halterman, R. L. *Tet. Lett.* **1989**, 30(27), 3513-16.

150. "Synthesis and Coordination Chemistry of 3a,7a-Azaborindenyl, a New Isoelectronic Analogue of the Indenyl Ligand." Ashe, A. J., III; Yang, H.; Fang, X.; Kampf, J.W. *Organometallics* **2002**, 21(22), 4578-4580.

151. Organic Syntheses: Collective Volume III: A Revised Edition of Annual Volumes 20-29. Horning, E.C. ed in chief. John Wiley & Sons, New York, NY. **1963**, p. 843.

152. "Chiral Cyclopentane-1,3-diyl-Bridged ansa-Titanocene Dichlorides" Chen, Zhuoliang; Halterman, Ronald L. *Organometallics* **1994**, 13(10), 3932-42.

153. (a) "Asymmetric Synthesis Catalyzed by Chiral Ferrocenylphosphine-Transition Metal Complexes. I. Preparation of Chiral Ferroenylphosphines" Hayashi, T.; Mise, T.; Fukushima, S.; Kagatoni, M.; Nagashima, N.; Hamada, Y.; Matsumoto, A.; Kawakami, S.; Konishi, M.; Yamamoto, K.; Kumada, M. *Bull. Chem. Soc. Jpn.* **1980**, 53, 1138-1151. (b) "New and Improved Synthesis of Opticaly Pure (R)- and (S)-2,2'-Dimethyl-1,1'-binaphthyl and Related Compounds" Maigrot, N.; Mazaleyrat, J. P. *Synthesis* **1985**, 317-320.

154. "Synthesis of ansa-2,2'-bis[(4,7-dimethylinden-1-yl)methyl]-1,1'-binaphthyl and ansa-2,2'-bis[(4,5,6,7-tetrahydroinden-1-yl)methyl]-1,1'-binaphthyltitanium and -

209

zirconium dichlorides" Halterman, R. L.; Combs, D.; Kihega, J.; Khan, M.A. *J. of Organomet. Chem.* **1996**, 520(1-2), 163-170.

155. (a) "Optisch aktive Binaphthyl derivate-Synthese Und Einsatz in Ubergangs metall katalysatoren" Brunner, H.; Goldbrunner, J. *Chem. Ber.* **1989**, 122, 2005. (b)"Cyclizations with hydrazine III. Syntheses of Pentaphene and Dinaptho [2,1-d: 1', 2'-f] [1,2] diazocine" Bacon, R.G.R.; Bunkhead, R. *J. Chem. Soc. Abstracts.*, **1963**, 839-845.

156. Halterman, R. L.; Morvant, M. C.; Philips, M. L. Organic Chemistry Laboratory Manual, fourth edition. RonJon Publishing: Denton, **1985**.

157. "Studies on the Diastereoselective Reduction of B-Hydroxy Ketones to 1,3-Diols With Common Hydride Reagents" Bonini, C.; Bianco, A; DiFabio, R.; Mecozzi, S.; Proposito, A.; Righi, G. *Gazz. Chim. Ital.,* **1991**, 121, 75-80.

158. "Chem Abstracts: Regio- and stereospecificity in propylene polymerization with chiral catalytic systems" Pino, P. Inst. Polym., Swiss Fed. Inst. Technol., Zurich, Switz. Editor(s): Lemstra, P. J.; Kleintjens, L. A. *Integr. Fundam. Polym. Sci. Technol.--2, [Proc. Int. Meet. Polym. Sci. Technol., Rolduc Polym. Meet.--2] (1988), Meeting Date 1987, 3-16.* Publisher: Elsevier, London, UK CODEN: 56IKA7 Conference written in English. AN 109:231615 AN 1988:631615 CAPLUS

159. Organic Chemistry Wade, L. G., Jr. Simon & Schuster: New Jersey, **1987**, p. 340.

160. "Preparation of Pure Enantiomers."

http://www.uis.edu/~trammell/organic/stereochemistry/resolution.htm

161. "Identification of Calcium-independent Phospholipase A_2 (iPL A_2) B, and Not iPL A_2 Y, as the mediator of Arginine Vasopressin-induced Arachidonic Acid Release in A-10 Smooth Muscle Cells" Jenkins, C.M.; Han, X.; Mancuso, D.J.; Gross, R.W. *J. Bio. Chem.* **2002**, 277(36), 32807-32814.

162. "The Horror and Hope of THALIDOMIDE"

http://colossus.chem.umass.edu/genchem/chem102/Articles/thalid.htm.

163. "An overview of chiral drugs" Rhodes, V. *Chimica Oggi/Chemistry Today* October **2002**, 20(10).

http://open.imshealth.com/webshop2/IMSinclude/i_article_20030123.asp.

164. (a) "Exciting times ahead for chiral technology" Frost; Sullivan. *Laboratorytalk*, www.laboratorytalk.com/news/fro/fro160.html. (b) "Stereoisomeric Drugs" Institute for International Research-USA. Conference Overview. www.iirusa.com/steroisomer. (c) "Single Enantiomer Drugs Keep Pace" Van Annum, P. **1999** Jan 18, Chemical Market Reporter. www.findarticles.com/m-FVP/3_255/53657145/p1/article.jhtml. (d) "Chirality Companies Broaden Their Approaches: Successful Firms Expand Beyond Single-Carbon Transformations" DePalma, A. *Genetic Engineering News* **2001**, 21(9) May 1.

165. "Olefin-polymerization with a single monomer" Abbas, R.; *Polymeric Materials Science and Engineering* **2001**, 84, 111.

166. "Crystal structure and propene polymerization characteristics of bridged zirconocene catalysts" Kaminsky, W.; Rabe, O.; Schauweinold, A.M., Kopf, H. *J. Organomet. Chem.* **1995** *(497) 181-193*.

211

167. "Degree of sterochemical control of racemic ethylenebis(indenyl)zirconium dichloride/methyl aluminoxane catalyst and properties of anisotactic polypropylenes." Reiger, B; Mu, X; Mallin, D.T.; Chien, J.C.W; Rausch, M.D. *Macromolecules* **1990**, 23(15), 3559-68.

168. "Homogeneous Ziegler-Natta catalysts. Part XI. Kinetics and stereochemical control of propylene polymerization initiated by ethylenebis(4,5,6,7-tetrahydro-1-indenyl) zirconium dichloride/methylaluminoxane catalyst" Chien, J.C.W.; Sugimoto, R. *J. of Polymer Science, Part A: Polymer Chemistry* **1991**, 29(4), 459-70.

169. "Cyclohexyl[*trans*-1,2-bis(1-indenyl)] zirkonium(IV) dichlorid: Ein chiraler Polymerisationskatalysator mit stereochemisch starrer Brucke" Rieger ,B. *J. Organomet. Chem.* **1992,** 428, C33-C36.

170. (a) "Crystal Structures and Solution Conformations of the Meso and Racemic Isomers of (Ethylenebis(1-indenyl))zirconium Dichloride". Piemontesi, F.; Camurati, I.; Resconi, L.; Balboni, D.; Sironi, A.; Moret, M.; Zeigler, R.; Piccolrovazzi, N. *Organometallics* **1995**, 14(3), 1256-66. (b) "Asymmetric functionalization of conformationally distinctive Cs-symmetric cis-[n.3.1] bicyclic ketones. Definition of the absolute course of enantio- and diastereodifferentiation." Underiner, T. L.; Paquette, L. A. *J. Org. Chem.* **1992,** 57(20), 5438-47. (c) "Conformational Cycloenantiomerism in 1,2-Bis(1-bromoethyl)-3,4,5,6-tetraisopropylbenzene" Singh, M. D.; Siegel, J.; Biall, S. L.; Mislow, K. *J. Am. Chem. Soc.* **1987**, 109, 3397-3402.

212

171. "Conformational Preferences of Racemic Ethylene-Bridged Bis(indenyl)-Type Zirconocenes: An ab initio Hartree–Fock Study" Linnolahti, M.; Pakkanen, T.A.; Leino, R.; Luttikhedde, H.J.G.; Willen, C.E.; Nasman, J.H. *Eur. J. Inorg. Chem.* **2001**, 2033-2040.

172. "Conformational Features of Group-4 *ansa*-Metallocenes with Long $-(CH_2)_n$ –Chains Connecting Their Cyclopentadienyl Ligands" Jodicke, T.; Menges, F.; Kehr, G.; Erker, G.; Howler, U.; Frohlich, R. *Eur. J. Inorg. Chem.* **2001**, 2097-2106.

173. "*ansa*-Metallocene Derivatives. 24. Deviations from C_2–Axial Symmetry in Ethano- and Etheno-Bridged Titanocene Complexes: Investigation of *ansa*-Metallocene Conformations" Burger, P.; Diebold, J.; Gutmann, S.; Hund, H.; Brintzinger, H.H. *Organometallics* **1992**, 11, 1319-1327.

174. (a)"Conformational Study of 2,2'-Homosubstituted 1,1'-Binaphthyls by Means of UV and CD Spectroscopy" Di Bari, L.; Pescitelli, G.; Salvadori, P. *J. Am. Chem. Soc.* **1999**, 121, 7998-8004. (b) "Synthesis of 1,1'-Binaphthyl-2,2'-dimethyl-Diyl-Bis(Inden-2-yl), A Chiral Homotopic c2-symmetrical Ligand and Its Derived Metal Complexes" Lotman, M. S. Masters Thesis, The University of Oklahoma. **1992**.

175. "Synthesis and Structure of C_2-Symmetric, Doubly Bridged Bis(indenyl)titanium and -zirconium Dichlorides" Halterman, R.L.; Tretyakov, A.; Combs, D.; Chang, J.; Khan, M.A. *Organometallics* **1997**, 16, 3333-3339.

176. "Synthesis of C7,C7'-Ethylene- and C7,C7'-Methylene-Bridged C_2-Symmetric

213

Bis(indenyl)zirconium and -titanium Dichlorides" Halterman, R.L.; Combs, D.; Khan, M.A. *Organometallics* **1998**, 17, 3900-3907.